T0317413

Automotive System Safety

Wiley Series in Quality & Reliability Engineering

Dr. Andre Kleyner

Series Editor

The Wiley Series in Quality & Reliability Engineering aims to provide a solid educational foundation for both practitioners and researchers in the Q&R field and to expand the reader's knowledge base to include the latest developments in this field. The series will provide a lasting and positive contribution to the teaching and practice of engineering.

The series coverage will contain, but is not exclusive to,

- Statistical methods
- Physics of failure
- Reliability modeling
- Functional safety
- Six-sigma methods
- Lead-free electronics
- Warranty analysis/management
- Risk and safety analysis

Published Titles

Prognostics and Health Management: A Practical Approach to Improving System Reliability Using Condition-Based Data
by Douglas Goodman, James P. Hofmeister, and Ferenc Szidarovszky
June 2019

Improving Product Reliability and Software Quality: Strategies, Tools, Process and Implementation, 2nd edition
by Mark A. Levin, Ted T. Kalal, and Jonathan Rodin
June 2019

Practical Applications of Bayesian Reliability
by Yan Liu and Athula I. Abeyratne
May 2019

Dynamic System Reliability: Modeling and Analysis of Dynamic and Dependent Behaviors
by Liudong Xing, Gregory Levitin, and Chaonan Wang
March 2019

Reliability Engineering and Services
by Tongdan Jin
March 2019

Design for Safety
by Louis J. Gullo and Jack Dixon
February 2018

Thermodynamic Degradation Science: Physics of Failure, Accelerated Testing, Fatigue, and Reliability Applications
by Alec Feinberg
October 2016

Next Generation HALT and HASS: Robust Design of Electronics and Systems
by Kirk A. Gray and John J. Paschkewitz
May 2016

Reliability and Risk Models: Setting Reliability Requirements, 2nd edition
by Michael Todinov
September 2015

Applied Reliability Engineering and Risk Analysis: Probabilistic Models and Statistical Inference
by Ilia B. Frenkel, Alex Karagrigoriou, Anatoly Lisnianski, and Andre V. Kleyner
September 2013

Design for Reliability
edited by Dev G. Raheja and Louis J. Gullo
July 2012

Effective FMEAs: Achieving Safe, Reliable, and Economical Products and Processes Using Failure Modes and Effects Analysis
by Carl Carlson
April 2012

Failure Analysis: A Practical Guide for Manufacturers of Electronic Components and Systems
by Marius Bazu and Titu Bajenescu
April 2011

Reliability Technology: Principles and Practice of Failure Prevention in Electronic Systems
by Norman Pascoe
April 2011

Improving Product Reliability: Strategies and Implementation
by Mark A. Levin and Ted T. Kalal
March 2003

Test Engineering: A Concise Guide to Cost-Effective Design, Development, and Manufacture
by Patrick O'Connor
April 2001

Integrated Circuit Failure Analysis: A Guide to Preparation Techniques
by Friedrich Beck
January 1998

Measurement and Calibration Requirements for Quality Assurance to ISO 9000
by Alan S. Morris
October 1997

Electronic Component Reliability: Fundamentals, Modeling, Evaluation, and Assurance
by Finn Jensen
November 1995

Automotive System Safety

Critical Considerations for Engineering and Effective Management

Joseph D. Miller

J.D. Miller Consulting, Inc.
MI, USA

This edition first published 2020

Registered Offices
John Wiley & Sons, Inc., 111 River Street, Hoboken, NJ 07030, USA
John Wiley & Sons Ltd, The Atrium, Southern Gate, Chichester, West Sussex, PO19 8SQ, UK

Editorial Office
The Atrium, Southern Gate, Chichester, West Sussex, PO19 8SQ, UK

For details of our global editorial offices, customer services, and more information about Wiley products visit us at www.wiley.com.

Wiley also publishes its books in a variety of electronic formats and by print-on-demand. Some content that appears in standard print versions of this book may not be available in other formats.

Library of Congress Cataloging-in-Publication Data Applied for

HB ISBN: 9781119579625

Cover Design: Wiley
Cover Image: © Chesky/Shutterstock

Set in 9.5/12.5pt STIX Two Text by SPi Global, Pondicherry, India

10 9 8 7 6 5 4 3 2 1

Contents

Series Editor's Foreword

The Wiley Series in Quality & Reliability Engineering aims to provide a solid educational foundation for researchers and practitioners in the field of dependability, which includes quality, reliability, and safety, and to expand the knowledge base by including the latest developments in these disciplines.

It is hard to overstate the effect of quality and reliability on system safety, a discipline covering engineering and management principles and techniques directed at identification, control, and reduction of the risk of harm. System safety expands to many industries including transportation, oil and gas, machinery, construction, aerospace, and many others. With over 30 000 deaths and over 200 000 serious injuries on the roads in one year in the United States alone, the automotive industry deserves special attention. There are no do-overs in safety. You cannot undo the damage to a human caused by an unsafe system; therefore, it is extremely important to design a safe product the first time.

Along with continuously increasing electronics content in vehicles, automotive systems are becoming more and more complex, with added functions and capabilities. The pursuit of autonomous vehicles intensifies this process, although with the hope of making vehicles safer and reducing the number of road injuries and fatalities. Needless to say, this trend is making the job of design engineers increasingly challenging, which is confirmed by the growing number of automotive safety recalls (for example, my four-year-old car has already had eight safety-related notifications). These recalls have prompted further strengthening of reliability and safety requirements and a rapid development of functional safety standards, such as IEC 61508 Electrical/Electronic/Programmable systems, ISO 26262 Road Vehicles and others, which have increased the pressure on improving the design processes and achieving ever-higher reliability as it applies to system safety.

The author of this book has been a long-time chairman of the International Society of Automotive Engineers (SAE) United States Technical Advisory Group (USTAG) to ISO TC22/SC32/WG08 Road Vehicle Functional Safety and is a recognized expert in the field of system safety. This committee has been one of the key contributors to the development of ISO 26262, both the original release and its 2018 upgrade, as well as ISO PAS 21448 concerning safety of the intended function (SOTIF) for driver assistance and vehicle automation safety. As a member of this committee, I had a firsthand experience and opportunity to appreciate the width and depth of Joe Miller's knowledge of system safety and engineering leadership skills.

This book covers various important aspects of safety including an overview of the implementation of the automotive functional safety standard ISO 26262, which consists of 12 parts totaling over 1000 pages. This material will help the reader navigate this comprehensive document, gain deeper knowledge of automotive systems safety, and make this book a perfect addition to the Wiley Series in Quality & Reliability Engineering.

We are confident that this book, as well as this entire book series, will continue Wiley's tradition of excellence in technical publishing and provide a lasting and positive contribution to the teaching and practice of quality, reliability, and safety engineering.

Dr. Andre Kleyner,
Editor of the Wiley Series in Quality & Reliability Engineering

Preface

If you are reading this book, you have made the decision based on your personal motivation to gain an expected benefit from the time and effort committed. The benefit is the acquisition of more practical information about automotive safety. The information you seek may be concerning a pragmatic approach to ensuring that safety is achieved for the products you support. You may be working on launch programs in the automotive industry with people who work in the safety domain and have to reconcile the need to only launch safe products with the need to always achieve the required launch schedule. This compelling need requires ethical resolution, and thus this book is relevant.

You are most likely aware of the high number of automotive safety issues covered by the news media. Such news coverage may raise a concern within you about how system safety can be ensured for automotive practice, and what the critical considerations are to ensure this safety. You may wish to gather information in order to make a judgment about what you can and should do. There are classes about liability, conferences on safety standards, and other informal sources that you may wish to supplement with practical considerations gained from my personal experience. With this background, you will feel ready to implement further improvements in process and practices in your workplace. Your informed initiatives will reduce product risk and thereby benefit the general public.

Work on popular standards such as ISO 26262 and ISO PAS 21448, leading to the future ISO 21448, is gaining a lot of publicity. You may be motivated to learn more about how these standards are used in a practical way. Insight as to why some requirements are included, and actual practice, will be useful to you. You might be an engineer with "standards fatigue" and be interested in getting the design done and the product launched without missing something that compliance to the standards could have prevented, but also without expending undue overhead to be compliant with quality and safety standards. Practical tips in my book may be welcome since I have actually been in charge of implementing an automotive safety process; helped write the standards; and worked in system, software, and hardware engineering as well as project management, program management, and operations. You can read when you can and get what you want from my book – your diligence will help launch safe products, and again, the general public will benefit.

You may be motivated to read this book because you are preparing to work in an automotive safety role such as safety engineer, safety manager, assessor, or auditor. You may have been assigned this role because of your general skill and potential or chosen the role because of your passion for safety. Either way, you want to make a difference through your skill and leadership. This requires ethical conduct and good judgment based on knowledge and experience. Learning through the experience of others by reading can accelerate this preparation. You want practical, pragmatic practices that you can consider, implement, and/or improve on you own. Just as important is knowing the advantages and disadvantages in order to avoid potential pitfalls while navigating the safety organization of your enterprise, or improving or establishing an effective safety organization. Perhaps your organization has been acquired or has acquired an organization with a different safety organization. Acting to ensure system safety diligence is needed and you are motivated to do it.

You may be experienced in another industry, perhaps as a safety expert in that industry. You have a different perspective than readers already working in the automotive industry: you are investing the time to read this book in order to learn about automotive practices in system safety, as well as critical considerations for the successful implementation of a system safety process in an automotive company. You may wish to compare and contrast the content of IEC 61508 with these automotive practices, as well as to learn about common practices of applying ISO 26262. If you have worked in transportation, such as rail or aviation, then there is a keen interest in automated vehicle safety. ISO PAS 21448 provides guidance for the safety of different levels of automation, so how this relates to ISO 26262 is of interest as well. With a strong background in system safety from other industries, this additional information will help the transition. You can then strongly support automotive safety.

You may be a consultant reading this book in order to acquire observations and judgment concerning implementation of a safety process. You may provide consulting services to organizations that have not yet established a formal system safety process but are growing rapidly and therefore need to ensure diligence with respect to system safety in current developments and establish a process to ensure safety for future developments. This is a critical service to provide, and you seek to broaden your basis for providing this service. My book may be selected because I have guided such a system safety project through process evolution for a multinational automotive supplier. Combining the content of my book with your experience will add value to your advice. Additional pitfalls can be avoided, improving execution.

You may be an enterprise executive who needs to ensure a best-in-class automotive safety organization to support the development and manufacture of the products sold by the enterprise. The safety organization may be splintered for numerous reasons. Or you may be a product line executive with similar needs or want to ensure an organization that meets the targets of the enterprise. You may be an executive of an engineering support group supporting several domains including safety, and these support groups may need to integrate into the enterprise to provide broad support. In all these cases, this book is directly applicable and discusses the pros and cons of different choices.

You may be an executive of program management responsible for managing the launch of safety-related products across multiple product lines and are concerned about establishing a process that integrates critical system safety milestones while maintaining schedule. This book provides relevant information. You may be responsible for engineering that includes safety, such as managing engineering domains that have assumed responsibility for the system safety process in addition to the engineering domain managed. Again, this book provides relevant information. You may be a program manager managing the development and launch of safety-related products and have concerns about how to achieve critical safety milestones while achieving the launch schedule – and this book provides relevant information. You are demonstrating diligence with respect to system safety, and this supports launching safe products.

You may have other motivations for reading this book. You may know me from encounters with the automotive safety standard; conferences where I presided, spoke, or was on a panel; classes I taught over the years; or business contacts. You may have a general interest in safety. You may be from a town in the British midlands, such as Nuneaton, and are curious about what a cheeky colonist like me might write in a book. You may be from a government agency involved with the automotive industry: information concerning safety provides background context for regulatory development and supports the system safety process.

I wrote this book to share experience-based judgments concerning system safety. These judgments are based on auditing and managing the safety-related work of automotive launch programs before there was an automotive safety standard, while supporting the development of standards, and thereafter. These programs introduced new technology to the general public. My experience also includes the successful development and deployment of a global system safety process, including the pitfalls and setbacks leading up to this success. Other background experience includes hardware, software, and systems engineering; operations; as well as military infrared and radar design, avionics, and radio communication. It is my intent to write about what may not be in the standards so that you can achieve the goals that motivate you to read and achieve personal success that benefits the general public. By reading, you may avoid the pitfalls and understand the methods that lead to success with the greatest efficiency. My practical tips will support this.

While my statements are my judgments based on experience, there is no guarantee of their accuracy or validity. You are responsible for the judgments you accept, and your actions. You may have different judgments or counterexamples: you may challenge or not agree based on your experience or opinions, and you may even find my statements offensive. You might accept the judgments and strongly agree; you may improve the recommendations and make them even more effective. The products you support will then have a lower risk when they are released; your processes will be deployed successfully.

I thank you for your passion and interest. I hope that your motivations for reading are satisfied and that you can improve safety. The act of reading this book demonstrates your energy and diligence, which are needed for this improvement. If the book is useful to you, then it is successful. You are responsible for your actions to improve safety. I hope that you will succeed – because then, the general public will benefit.

Abbreviations

ADAS advanced driver assistance system
AI artificial intelligence
ALARP as low as reasonably practicable
ASIC application specific integrated circuit
ASIL automotive safety integrity level
C controllability
CCS calculus of communicating systems
CSP communicating sequential processes
CORE controlled requirements expression
DFA dependent-failure analysis
DFI dependent-failure initiator
DIA development interface agreement
E exposure
ECC error-correcting codes
EOTTI emergency operation tolerance time interval
EQA engineering quality assurance
ESD electrostatic discharge
FMEA failure modes and effects analysis
FMECA failure modes and effects criticality analysis
FTA fault tree analysis
FTTI fault-tolerant time interval
GES general estimates system
HARA hazard and risk analysis
HMI human machine interface
HOL higher-order logic
IMDS International Material Data System
IP intellectual property
JSD Jackson system development
LFM latent fault metric
LOPA layer of protection analysis
MC/DC modified condition/decision coverage
MEMS micro-electro-mechanical systems
MSIL motorcycle safety integrity level

MTTR	mean time to restoration
NASA	National Aeronautics and Space Administration
NASS	National Automotive Sampling System
NHTSA	National Highway Traffic Safety Administration
NM	Newton-meter
OEM	original equipment manufacturer
OTA	over the air
PAS	publicly available specification
PFMEA	process failure modes and effects analysis
PHA	preliminary hazard analysis
PIU	proven in use
PLD	programmable logic device
PMHF	probabilistic metric for random hardware failures
RfQ	request for quotation
S	severity
SEooC	safety element out of context
SIL	safety integrity level
SOC	system on a chip
SOTIF	safety of the intended function
SPFM	single-point failure metric
T&B	truck & bus
VM	vehicle manufacturer

1

Safety Expectations for Consumers, OEMs, and Tier 1 Suppliers

Every business involves sales based on customers trusting that their expectations will be satisfied. This trust by the customer is based on a belief that ethical business practices are the norm and not the exception. The customer expects to be satisfied and receive value for money. No one expects to buy disappointment, and the global automotive business is no different in this regard. To some extent, the belief in ethical business practices is supported by the observance of regulatory enforcement. Media coverage of investigations and recalls adds credibility to this belief. Consumers believe that being treated ethically is a right, not a privilege. There are consumer expectations of performance, prestige, and utility. Part of this utility is trustworthiness.

Trustworthiness

Trustworthiness includes quality, reliability, security, privacy, and safety, and expectations of trustworthiness are increasing. Advocates of quality publish competitive results for quality among competing automotive vehicle suppliers. Competing vehicle suppliers reference these publications in marketing campaigns targeting consumers' belief that they are entitled to the highest quality for their investment in an automobile. Fewer "bugs" in new cars are expected; taking the vehicle back to the dealer in two weeks due to an initial quality defect is no longer acceptable to automotive consumers. The consumer expects improved reliability to be demonstrated by longer life with fewer repairs. Consumers review warranties for length and coverage as a way to improve their confidence that potential future maintenance expense will be manageable. Comparisons are made of warranty repairs to support purchasing decisions. Consumer advocates track reliability for this purpose; it is a product differentiator.

When they purchase a vehicle, consumers expect that the vehicle will be secure and robust against cyber attacks. No loss of service availability is expected due to hacking of vehicle systems. Security of entry mechanisms is expected to be robust against cyber attacks by potential thieves. Even though successful hacks are in the news, it is expected that security considerations sufficient for lifetime protection are included in the vehicle design. If updates are needed, no intrusion or loss of availability is acceptable. Privacy of personal information is expected to be protected, even if some information is used to provide vehicle

Automotive System Safety: Critical Considerations for Engineering and Effective Management, First Edition. Joseph D. Miller.

service. Ethical treatment of data concerning driving behavior, locations visited, and frequently used routes is expected. Permission to share this data may not be assumed. Privacy is considered a right, and enforcement is expected.

Perhaps most important are expectations of safety. Even when advanced convenience features are included in a vehicle, the consumer expects that there is no additional risk. The risk of harm is expected to be the same or less than it was before the features were added. When consumers acquire and operate an automobile, they are demonstrating an acceptance of the risk of operating the vehicle based on the current state of the art, and this state-of-the-art assumption does not include an assumption of increased risk. Consumers may not cognitively be accepting a state-of-the-art risk – they may not even be aware of what the state-of the-art risk is at the time of purchase. Nevertheless, they accept it by their actions. They purchase the vehicle knowing that there is a risk of harm when operating a road vehicle. This indicates accepting state-of-the-art risk.

Consumers have an expectation of safety. What is this expectation? How can risk be determined to be consistent with the state of the art at the time of purchase? This determination depends on the definition of *safety* and how that definition is employed in automotive practice. There are several candidate definitions to consider. *Safety* has been discussed as meaning "no accidents." This is aspirational; consumers are expected to welcome the freedom to operate a vehicle without the risk of accidents, especially if such freedom from accidents can be achieved at a reasonable cost. While useful for some analysis [1], the current state of the art for automotive vehicles has not yet advanced to this stage. Convenience features are being added to move in this direction, and vehicle manufactures reaffirm such shared aspirations in their marketing campaigns. The news reports on progress and setbacks along the journey to reach this aspirational goal of automotive technology. Clearly it has not been achieved – it is not yet the state of the art for automotive safety. Still, consumers purchase vehicles knowing there is a risk that they may have an accident and die while driving a vehicle, and they drive anyway. They accept this risk by their actions. Consumers know there is a risk of death to their loved ones who travel in the vehicle they purchase. Still, they drive their loved ones – they have accepted the risk.

Another definition of *safety* is "absence of unacceptable risk" [2]. The definition may be applied in any situation or scenario, whether related to the automotive industry or not. In this definition, safety is not absolute; the concept of risk is introduced. *Risk* is defined as the combination of the probability of harm and the severity of that harm. *Harm* is defined as damage to persons or, more broadly, as damage to persons or property. *Acceptable* in this context is ambiguous in that it implies that someone defines what risk is acceptable.

A similar definition is the "absence of unreasonable risk" [3]. This definition is also used in non-automotive scenarios or applications. However, it is the definition chosen for the functional safety standard used in the automotive industry. It seems reasonable to conclude that the consumer accepts that the risk is not unreasonable if they purchase or drive the automobile. While the consumer prefers the risk did not exist at all and that there were never any automotive accidents, by their actions they have shown that they consider the risk not unreasonable when considering the benefits provided by driving. This is the basis for the automotive functional safety standard, ISO 26262.

To not be unreasonable, the risk must not violate the moral norms of society. These moral norms may change over time, as do the expectations of consumers. However, the norms of society are not aspirational. Recalls occur if it is discovered that these norms might be violated due to an issue with a vehicle. Since the current rates of various accidents, including fatal accidents, are the norms of society at the time a vehicle is purchased, they define reasonable risk for the consumer. Consumers decide the risk is not unreasonable when they make the decision to purchase a vehicle. The consumer does not expect that the vehicle will put people at a greater risk than other cars already on the road do – the consumer does not expect to purchase a defective or an inherently unsafe vehicle. Rather, the consumer expects the risk to be the same or less, depending on information they have received through the media or dealership concerning the new state-of-the-art features that are included. Even new features that do not have data to confirm their safety, from years of experience on public roads, are expected to improve, not diminish, the state of the art with respect to safety. Consumers consider this not unreasonable. They consider this safe.

Consumer Expectations

The consumer may choose to purchase a vehicle differentiated by advanced driver-assistance features not yet included on all the vehicles in service. Even if these advanced driver-assistance features are available for vehicles on the road, there may not be sufficient data to determine the risk to society of these features. Improvements and additions may be made to these features, or there may be more interactions of these features with other automotive systems that have the potential to cause harm. Now the expectations become less clear because they are not based on data that includes the influence of these advanced systems.

Expectations are influenced by advertising, news data about similar features, and personal experience. There has been much in the news about automated driving without these vehicles being broadly available to consumers. Still, expectations are being influenced by information provided by the media. Media reports of fleets of automated vehicles raise awareness of the many successful test miles as well as any errors or accidents that are publicly reported. This information may raise or lower expectations of advanced driver-assistance features that have some similarity to automated features. Automated vehicles have control of steering systems and braking systems in a manner similar to emergency braking and lane-keeping assist systems. Some clarifications might be discerned from the media, which explains the expectations of driver responsibilities and awareness. Expectations of the automated system are clarified; the capability of assistance has limitations.

Nevertheless, such publicity can have the effect of increasing consumer expectations regarding the performance of advanced driver assistance systems (ADAS) that are available. The more publicity there is about automated driving successes, failures, improvements, and goals, the greater the anticipation of its availability. The anticipation of available automated driving features may distort the understanding of the capability of more-limited features. This has been discussed as leading to a cyclic variation in consumer expectations based on experience. The consumer may not fully appreciate the nuanced limitations of an ADAS.

For example, the consumer may initially have high performance expectations for a follow-to-stop radar cruise control system. The consumer expects the system to perform as the driver would perform in all circumstances. This is reinforced by the early experience of having the vehicle slow to a stop automatically while following another vehicle. Gradually this experience leads the driver to not hover a foot over the brake pedal: the driver observes but does not intervene, and confidence starts to build. Then the driver mistakenly expects the Doppler radar cruise control system to stop for a vehicle that was not being followed and that is sitting still at a traffic light. This is consistent with the behavior expected from a human driver. However, the ADAS does not respond because it ignores stationary objects in its field of view that are not being followed, like bridges and trees. This is a technical limitation of the system but is consistent with the requirements of the design. The consumer's mistaken expectations are not satisfied, and the consumer's opinion of the product becomes less favorable.

The expectations of the consumer are not consistent with the required capability of the ADAS. Further, the system sometimes mistakenly brakes when the vehicle in front slows and then changes lanes. The vehicle in front appears to disappear, consistent with the appearance of a stopping vehicle. Consumer sentiment drops further.

The consumer's experience continues with more successful following, acceleration, and proper behavior when changing lanes. In these scenarios, the ADAS reacts in much the same way as a human is expected to react. The consumer's expectations now are being calibrated. The consumer is ready to intervene when the ADAS's limitations are exceeded but does not intervene when the system is capable of handling situations successfully. Overall, the consumer does not feel the risk is unreasonable. The ADAS is not an automated driver, and it will not handle every situation like a human driver. However, it is pretty good; it seems safe.

Unless the consumer has confidence that their expectations of safety are satisfied by a vehicle, they will not purchase that vehicle. This confidence may be influenced by publicity, publicly reported performance and performance comparisons, and word of mouth. It is clear that consumer expectations of safety are critical to the automotive business, because the safety concerns of potential customers can severely limit sales of a vehicle. Tremendous resources are deployed not only to influence these expectations, but, more importantly, to satisfy them.

Vehicle manufacturers (VMs) expend resources to promote the advances they have made to improve crashworthiness. They spend vast resources to continuously improve the crashworthiness of the vehicles they intend to manufacture. Resources are provided to support development of improved passive safety systems to protect occupants and pedestrians during an accident. Included in these resources are not only provisions for development engineers, but also resources for engineering quality, safety management, and execution of the safety process. These resources are deployed by both the original equipment manufacturer (OEM) and suppliers. Each has its own process and resources to ensure the safety of its own products. They may share and coordinate resources for joint development of safety-related systems; safety resources must be managed effectively in both individual and joint developments. Effective management of a safety organization is discussed in Chapter 2, where evaluation criteria and alternative organizations are evaluated that may be considered equally by OEMs and suppliers. Both the OEM and the supplier have safety requirements to define and must comply with these safety requirements. Regulatory agencies and customers expect this – these expectations must be satisfied.

OEM Expectations

Safety expectations for the OEM or VM go beyond the vehicle not placing the consumer at unreasonable risk. This is only the minimum requirement for a safe vehicle as defined. Evidence of fulfilling this minimum requirement supports confirmation of diligence by the OEM. The OEM accepts responsibility for this expectation and seeks in addition to differentiate the vehicle in the marketplace by further reducing the risk to the consumer. The VM strives to attain this goal by providing additional resources for continuous improvement in the performance of the safety-related content of the vehicle. This is a resource-consuming and challenging task. Research can be performed to identify areas of opportunity where resources may best be deployed. Realistic goals are established and promoted as future capabilities for the enterprise; these goals can then be used to derive specific objectives for product development engineering to achieve within a specified time period. This allows development to be planned for and resourced, taking into account both the specified time and the process used by the enterprise to ensure that the safety requirements are identified so compliance will be achieved.

Third parties independently rate vehicle "safety" by measuring crashworthiness as well as performance of various ADAS featured on the vehicle. Crashworthiness is evaluated using specific tests to determine the safety margin achieved by the design under repeatable conditions, as well as the absence of potential hazards previously determined, such as ensuring the safety of infants in car seats. ADAS features evaluated include emergency braking for both forward driving and backing the vehicle.

OEMs develop requirements for the vehicle design, specify features to be included, and identify requirements for these features. System safety plays a major part in determining these requirements. For example, it may be determined that to improve safety-related features, the vehicle will contain a safety feature that detects pedestrians approaching the front of the stopped or moving vehicle and that automatically stops the vehicle to avoid a collision with the pedestrian. Requirements at the feature level include the proximity of the pedestrian, scenarios in which the vehicle will be stopped, scenarios in which the vehicle will not be stopped, and a degradation strategy in case of failures. These requirements may be developed by the OEM, jointly with suppliers, or by the suppliers themselves. Arbitration of the priority of a vehicle request from this feature compared to requests from a stop-and-go cruise control or the driver are resolved between the VM and suppliers. Integration of the features may be executed by the OEM or delegated to a "turnkey" capable supplier. Such a supplier may supply multiple safety-related systems, or have the capability to supply other safety-related systems, and can employ this detailed domain knowledge to determine a safety policy for arbitration. Such a policy may require validation as described in ISO PAS 21448. Then the VM supports the validation plan. The VM assumes ultimate responsibility.

The OEM expects evidence that these tasks are complete and ensure safety, e.g. a safety case. This evidence includes proof that the relevant standards have been satisfied by the process employed to complete the tasks. There should be evidence that sufficient analyses have been performed to provide confidence that the requirements are complete because of the systematic elicitation of requirements through safety analyses. The OEM also expects the vehicle to meet regulations applicable to the performance of included systems. For example, if a stability control system is included, then regulations concerning stability control for each nation in which the vehicle is to be offered must be met. There may be regulations concerning ADAS that also need to be met. Meeting these OEM safety expectations helps ensure that consumer expectations are met.

The consumer may not be aware of the systematic methods employed to elicit and verify safety requirements; they also may not be aware of the tests performed to prove that the applicable regulations have been met (although some testing may be presented in advertisements). Nevertheless, these systematic methods help ensure that only safe, fault-free systems are released to the general public. Success in meeting these safety expectations can be achieved by completely determining the safety requirements and completely complying with them. Requirements must be met for the entire vehicle life cycle: the concept phase, design phase, and verification phase, as well production and safety in use, in repair, and even in disposal. The safety lifecycle will be discussed in Chapter 6; it can be tailored for each project, and the scope may change.

Supplier Expectations

Tier 1 suppliers – suppliers providing systems directly to the OEM – expect to provide systems that meet the expectations of the OEM, if the OEM meets the underlying assumptions that are the basis of the design requirements for these systems. For example, a tier 1 supplier may assume that the OEM will limit the braking command of a cruise control system to a safe level in the braking or engine management system. The tier 1 supplier may also assume that the OEM checks messages for transmission errors and takes a safe action. These assumptions are necessary because tier 1 suppliers must develop their technologies in anticipation of future OEM expectations. Such advanced technology development by the supplier may take much longer that the time allowed by the OEM for sourcing systems for a vehicle to be launched in a specified model year. Tier 1 suppliers' scheduling must take into account the lead time to develop the baseline technology needed for the system to be provided by the supplier.

For example, a supplier of a radar cruise control system must develop the essential product technology before the supplier can support a vehicle launch schedule for an OEM. Antenna technology for the frequency band and field of view require extended development time. Encoding the radar transmission to support specific detection goals also requires basic development to be suitable for production intent concepts. The supplier must perform safety analyses to determine the safety requirements, and evidence is required for compliance with these requirements. Compliance with some requirements is assumed based on assumptions for the OEM, such as vehicle behavior in case the system indicates an internal fault, mounting alignment, and an unobstructed field of view. The OEM must confirm these assumptions, or equivalent assumptions and measures need to be agreed on before sourcing. For example, external detection of a faulty system by another tier 1 supplier may be managed by the OEM. These assumptions by the supplier become requirements for the OEM. Evidence of verification is needed for the OEM safety case; this evidence is traceable to the assumption, and compliance is expected.

Further expectations of suppliers and OEMs are supported by standards such as ISO 26262. This standard provides requirements for the roles of the supplier and the customer that support the joint development of safety-related systems. Sharing safety analyses, the resulting safety-related requirements, and other work products is specified for joint development. Collaboration by OEMs and suppliers while creating standards helps to vet the standard and support completeness of the requirements with respect to the standard's scope. OEMs from many nations participate in determining these requirements with tier 1 and tier 2 suppliers internationally. In the case of ISO 26262, the scope is limited to road vehicle functional safety.

Functional safety is the safety of the system as related to functional failures. The difference between this and system safety will be discussed in Chapter 3. Nevertheless, the standards are intended to provide a common language and frame of reference for OEMs and suppliers. This frame of reference includes evidence of compliance with standard requirements, such as work products. OEMs and suppliers agreed in ISO 26262 to have a development interface agreement (DIA) to structure the work and exchange of information concerning functional safety. The DIA is framed in the context of customer expectations, which will also shape the requirements for system safety and influence the expectations of both parties of the DIA.

Consider Figure 1.1, which illustrates critical safety considerations that could be easily overlooked. Consumer expectations for a new vehicle are top center and the highest

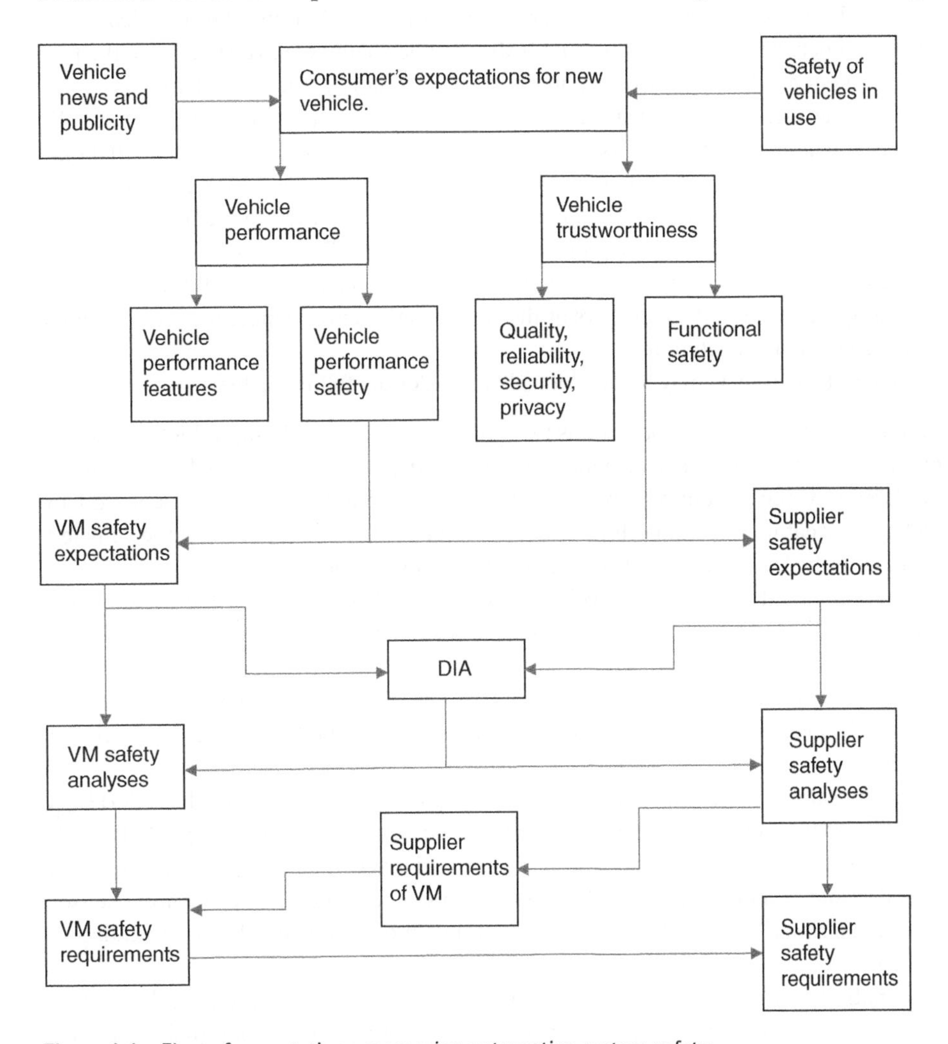

Figure 1.1 Flow of expectations concerning automotive system safety.

priority. These expectations of improved vehicle safety only contribute to the safety of vehicles in the field if the OEM and suppliers satisfy those expectations sufficiently that the consumer purchases the vehicle. Therefore, the influence of information about vehicles in the field as well as from publicity is a logical driver of these expectations. Often, advertisements demonstrate the advantages of new active safety features: for example, an accident is avoided because of an action taken by the advanced safety system. On the other hand, the news may report the failure of an autonomous system in a particular scenario, as well as the overall reduction in accidents due to advancements in safety systems.

Figure 1.1 shows that expectations of consumers include input to the vehicle's performance and functional safety. These expectations are made known to the VMs and suppliers through buying preferences for advanced systems of one VM over those of another VM. Information published about reliability and privacy as well as security and quality also influences consumers' purchasing preferences. These preferences are a fundamental input that shapes the goals that VMs and suppliers expect to achieve. This input is included in the collaboration between tier 1 suppliers and VMs. Suppliers compete to differentiate the features of the systems they supply based on their ability to satisfy consumer demands.

Shortcomings identified by consumer feedback to VMs are also acquired by tier 1 suppliers and used to guide improvements in subsequent systems. This fosters greater collaboration between VMs and suppliers. In order for this collaboration to meet the expectation that the vehicle is as safe as vehicles in the field, the standards that provide guidance to achieve the safety of vehicles in the field must be taken into account. Figure 1.1 shows the DIA as referenced in ISO 26262. It provides guidance for supplier selection as well the roles each collaborating party plays. All members of the relationship strive to meet the requirements and meet consumer expectations.

So, what critical aspects may be overlooked? Consider the following list:

1) Calibration of consumer expectations for vehicle features considering vehicle news and publicity. Limitations of performance and availability are communicated: for example, availability of the system may be limited, or it may be automatically switched off in inclement weather. System performance may not be sufficient to maintain the vehicle's position in the lane during high-speed or sharp-radius curves.
2) Determining field data related to vehicle features and ensuring that safety is not worse than that in the field. Accident data is a source of such information and can be obtained, for example, from the National Highway Traffic Safety Administration (NHTSA) National Automotive Sampling System (NASS) General Estimates System (GES). This data can be used to determine the risk of certain accidents and the severity of injuries in the field.
3) Including safety of the intended function (SOTIF) in the DIA or equivalent. SOTIF is out of scope for ISO 26262; ISO PAS 21448 [4] discusses SOTIF. From this publicly available specification (PAS) requirement, tasks can be derived and assigned to each party to establish and achieve the goals of the PAS. These could include determining sensor limitations as well as establishing and executing a model and vehicle-based validation strategy.
4) Systematic safety analyses by both the VM and supplier, e.g. architecture and software, and sharing the assumed requirements with each other. These may include failure

detection and performance during a failure. The analyses may include a hazards and operability study (HAZOP) of the system architecture as well as the software architecture so the parties can systematically agree on a failure-degradation strategy. Modifications may result from the analyses.

These are critical considerations because they relate to safety requirements. Considering these demonstrates diligence with respect to protecting consumers and other users of public roads.

These requirements all flow from consumer expectations but are easily overlooked because they appear to come from independent sources. Customer expectations, accident databases, independent standards, and systematic safety analyses may seem to be unrelated. For example, an engineer may look at customer requirements and assume that they are complete. The customer requirements are extensive and cover performance requirements, the product concept, and perhaps some design-specific details. Verification requirements and validation requirements are also addressed. Eliciting requirements from systematic software architectural analyses is not required to comply with these extensive customer requirements. Nevertheless, to meet societal norms, such analysis of systems in the field is required, to avoid unreasonable risk. Similar systems in the field may have had such systematic analyses performed to increase confidence that the requirements are complete. The analyses demonstrate that actions were taken to ensure compliance with requirements; this is a reasonable expectation. Requirements, elicitation, and management will be discussed in Chapter 8.

Engineers install failsafe safety mechanisms as a safety system is designed and not as a result of systematic analyses. The original system concept may include safety mechanisms that are deemed appropriate; such safety mechanisms are at different levels and provide redundant protection. Good engineering judgment is exercised, but not in the context of what is expected as described previously. It can be difficult to show that all required safety mechanisms have been included: experience, rather than systematic analyses, may be a basis for this judgment, and the measures may have sound reasoning. Faults are inserted, and the correct operation of the safety mechanism is demonstrated. This verification of the safety mechanism is traceable to the safety mechanism requirement. Nevertheless, the evidence of completeness illustrated in Figure 1.1 is not available. It can be difficult to support with high confidence an assertion that there are no missing requirements. Such confidence of completeness comes from a structured requirements elicitation method rather than judgment alone. Systematic safety analysis is an accepted method to perform this elicitation: this is a critical consideration for system safety.

2

Safety Organizations

The Need for a System Safety Organization

Meeting consumer expectations for system safety in automotive products requires diligence to ensure that no unreasonable risk occurs after consumers purchase a product. There must be no unreasonable risk to either the consumer or to others exposed to the product's use throughout the product's life. The intended use and foreseeable misuse of the product must not cause unreasonable risk, and there also must be no unreasonable risk imposed on the consumer or others as a result of a failure of the product. This diligence must be driven consistently throughout the product organization at the vehicle level and at the supplier. Consumers expect a very intense effort to ensure their safety and the safety of their families by all who are involved in developing, manufacturing, delivering, and repairing the product. This includes executive management and functional management in both engineering and operations, including support functions.

Figure 1.1 shows what must be fulfilled in order to achieve what is expected for safety. As discussed in Chapter 1, these expectations can be easily overlooked. Thousands of engineering and operations personnel are involved in the launch of an automotive product, and diligence is required from all of these personnel to fulfill customer expectations for safety. Considering the large consumption of resources required to launch a road vehicle, as well as resources consumed developing and launching each of the vehicle's systems, a systematic approach to diligence with respect to safety is appropriate. To deploy this systematic approach across the enterprise requires technical support for execution. It requires management of the process, which also requires resources.

Consider a global tier 1 safety system supplier that supplies advanced driver assistance systems; braking, steering, and passive restraint systems, as well as other products with safety requirements. Such a supplier has multiple engineering development centers around the world. These development centers can support collaboration with local and regional vehicle manufacturers as well as global vehicle manufacturers VMs that have local development centers in the region.

There are cultural differences from development center to development center as well as multiple engineering organizations within each development center. The cultural differences may be due to different languages and practices for everyday living and social

Automotive System Safety: Critical Considerations for Engineering and Effective Management, First Edition. Joseph D. Miller.

interactions, as well as differences in business practices. Customer expectations are communicated by the customer in a more formal or less formal way from one technical center location to another. Schedules and content for the vehicle launch may be more carefully planned in advance or modified in an almost ad hoc manner throughout the launch cycle.

The engineering organization at a technical center may be organized by engineering specialty, such as software engineering, electrical hardware engineering, mechanical engineering, and system engineering; or by product function, such as applications engineering or advanced product development independent of the particular application. Both applications engineering and advanced development engineering may also be organized by engineering discipline. There may be an engineering quality organization that checks a sample of the projects to determine if the process steps are performed on time. A global quality standard process can be instantiated on each project being supported by the engineering quality department in order to ensure compliance with the engineering quality standard and to standardize on a global engineering process for the enterprise.

Each of these engineering specialties has particular capabilities for design and analyses. System engineering provides the concept and high-level requirements for the system, hardware, and software. Software engineering derives software requirements for the software architecture, software modules, and software units. Hardware engineering develops requirements and design for electrical circuits. However, none of these engineering specialties is system safety. The specialized elicitation and derivation of safety requirements is not the specific domain of system, hardware, and software engineers; nor is the establishment and execution of a safety process included in the domain of engineering quality. How are system safety tasks managed? How are they executed diligently?

Many – perhaps most – automotive suppliers and VMs have safety organizations. The need to establish, support, and execute a safety process is universally accepted across the automotive industry. This goes beyond what is legally required by the regulations of each country, or the performance requirements in safety system regulations. These safety organizations may be separated into product safety and functional safety, or they may be combined into a system safety organization.

The differences in these approaches are somewhat regional: for example, in the United States, it is common to see system safety organizations that include the safety of systems in case of failure and also consider the safety of the systems' intended functions without any failures. For example, such an organization considers the safety of an automatic emergency braking system in the case of a false detection of an object while the automatic emergency braking system is operating within its specification. In Germany, the safety of the intended function is addressed by the product safety organization, and the safety of systems in case of failure is addressed by the functional safety organization. System safety versus functional safety in automotive applications will be discussed in Chapter 3.

Functions of a Safety Organization

These safety organizations support the achievement of system safety throughout the safety life cycle of the product. This life cycle includes the concept phase, when the overall functional concept of the product is established and analyzed for potential hazards; the design phase; the verification phase; release for production; use; repair; and disposal. Activities include:

1) Determining or recommending a global safety policy for the supplier or VM. Such a safety policy directs the enterprise's general goals with respect to safety and references where process documentation responsibility may be found, or the policy may include specific responsibilities for safety-related tasks.
2) Developing and maintaining a system safety process. The safety process will detail the specific safety-related tasks, management of execution of these tasks, and responsibilities for execution, assessment, and auditing.
3) Executing systematic analyses of the system, hardware, and software. Many of these systematic analyses take the form of a HAZOP, such as a hazard and risk assessment of the system functions. Other analyses include a software architectural analysis, hardware failure modes and effects analysis (FMEA), and single-point fault analysis.
4) Training system, hardware, and software engineers as well as management, operations, and supporting functions concerning their roles in achieving system safety. This training may include basic online process training, specialized training for each engineering domain, operations training including background for interaction with outside auditors, overall priorities, expectations of management, how achievement of these expectations will be reported to executive management, and executive actions that are required.
5) Assessing work products. This includes what level of independence is required for assessment of each potential work product, what tools are available for performing these assessments, communication between the author and the assessor, how long it may take, what should be included in the assessment, required actions on assessment findings, storage of the assessment, release of the document, and what may be shared with the customer or outside parties.
6) Auditing projects and reporting. This includes periodic audits of all projects by a specified auditor, actions and follow-up on all actions, reporting the audit report, actions still outstanding, and any escalations to senior management for urgent resolution.

To perform these activities efficiently, a safety organization is established and staffed. The staff must include the necessary talent to execute each of activities just described. Assessments and auditing need to be performed by staff that is independent of the release organization. The safety analyses need to be performed by individuals with domain knowledge and associated with the release authority.

This organization may be located anywhere in the automotive supplier's or VM's organizational structure, as long as the necessary independence is maintained for assessment and auditing. Staff for other activities, such as training and performance of analyses, may also be independent but are not required to be. However, there are advantages and disadvantages in how the safety organization is organized and where it is placed. There are critical success factors to consider, which will be discussed next.

Critical Criteria for Organizational Success

There are five critical success factors to consider for a system safety organization [5]:

1) The organization must have the talent to perform the safety tasks.
2) System safety must be integral to product engineering.

3) There must be a career path for safety personnel, to retain them in a safety role.
4) The safety process must be owned by program management so that the tasks can be planned, resourced, and executed.
5) There needs to be a periodic executive review cadence to ensure that the process is followed.

Unfortunately, unless all five critical success factors are met, the system safety organization will fail. It is possible to organize safety within an enterprise while not achieving these critical success factors. It is also possible execute product development and launch while not considering system safety for safety-related products. This is a failure of the safety organization – and the enterprise itself may fail.

Talent to Perform the Safety Tasks

Failure to acquire and retain the necessary talent to perform the tasks may cause analyses to be missing, misleading, incomplete, or in error. The documents themselves may be written in such a way that required actions are not clear. The documents may be ambiguous and adversely affect the safety case because they do not communicate with clarity what the findings are, how to resolve them, and the correct status of the resolution. Thus, an action may be raised but not documented when closed. If this is not discovered, unreasonable risk may occur due to missing or erroneous safety requirements for the automotive product being developed. If the errors are discovered, the system safety organization loses credibility and becomes less effective within the VM or automotive supplier. The safety organization may itself be seen as a potential liability to the enterprise and be avoided. Either way, the needed trust is lost, and the organization fails.

Integral to Product Engineering

System safety also needs to be an integral part of product engineering. Ideally, system safety specialists are co-located with product engineering and participate in all communications. Changes and nuances in the product's design and function should be deeply understood and considered in all safety analyses. If the system safety organization performs tasks independently of product engineering, the basis of the analyses may be flawed. The actual effect of a failure may be misunderstood. The specified performance may not be accurately considered when evaluating the safety of the intended function: it might be based on assumptions about the product design that are not consistent with the concept of product engineering. Because of this independence of the safety analysts from the product engineering personnel being supported, and the resulting ineffective communication, these misconceptions may not be resolved.

The requirements outlined earlier may be flawed not due to incompetence, but due to poor communication with the designers. Even if the assumptions are correct, the implementation of safety mechanisms and measures may not be confirmed. The safety experts may not have the ready access necessary to consider verification methods or absence of verification. Verification may or may not successfully show compliance. There is evidence of unsatisfied requirements, and the organization fails.

Career Path for Safety Personnel

All around the world, system safety professionals are interested not only in practicing their profession, but also in advancing their careers. They may not be content to continue in a role indefinitely with no opportunity to broaden their technical expertise or assume new responsibilities. In order to retain safety professionals with the talent needed for the VM or automotive supplier's products, the safety organization must take into account the professionals' desire for career advancement. The value of each safety expert in the organization grows rapidly with experience and product knowledge. The experience and product knowledge of these safety experts strengthens the enterprise and can be shared with junior safety personnel as well as non-safety personnel responsible for tasks in development or operations.

The career path should be in the safety organization, if system safety skills are to be improved. Continuous mentoring can occur in the safety organization; continuity can be ensured for the enterprise using this strategy. If a career path is provided, but not within the system safety organization, then high-potential safety professionals will leave system safety to pursue a career outside of safety. Safety then becomes background knowledge that the candidate has acquired and may use in their next job. Organizationally, this drives a continuing gap in the talent required to perform, audit, and assess the tasks that ensure no unreasonable risk to the general public. This is not a sustainable environment for the safety organization or the enterprise.

Similarly, if there is no career path, safety professionals may seek to advance their careers in system safety at a competitive VM or automotive supplier. The enterprise then becomes a training ground and steppingstone to be used by safety experts to leave and advance their careers. This can strengthen the receiving competitor's system safety organization: the competitor gains not only the safety expert's knowledge of system safety, but also knowledge of the strengths and vulnerabilities of their competitor. However, it will systematically weaken the donor. That organization fails.

Safety Process Owned by Program Management

It is possible to launch an automotive product without properly performing all system safety tasks on schedule. Priorities may encourage a launch team to delay safety tasks in order to meet other early milestones in the program. Resources that are needed to complete early safety activities may be assigned to tasks to meet milestones agreed on with the customer. The same resources may be over-assigned to the safety activities because of an overall shortage of resources as the program ramps up. And not performing safety tasks may not physically prevent product development and manufacturing: for example, if the hazard and risk assessment is not completed, the concept samples can still be produced and delivered to the customer.

Safety tasks are now considered the norm for the industry. VMs expect that the basic activities of each phase of the safety life cycle are completed on time within the appropriate phase. Consumers expect that normal safety tasks are completed prior to the product being offered for sale, because unreasonable risk may otherwise result. For example, failure to identify all potential hazards using a hazard and risk analysis may result in the requirements not being identified to avoid or mitigate that hazard. The requirements may not be satisfied.

Ensuring that normal safety tasks are performed on time requires that they be planned, resourced, and executed in the launch program prior to manufacturing. This must be accomplished with sufficient priority so that resources are not reassigned due to conflicting priorities. This planning of safety activities therefore must be ensured by including these tasks with those owned by program management.

For an enterprise to successfully launch products, it must have an organization that successfully manages programs. This implies that priorities are maintained, and execution is monitored and ensured. The tasks can be executed even if they are not included in the tasks owned by program management, but resourcing and execution may not be ensured. Conflicting priorities could divert resources due to immediate needs; or there could be conflicts for resources if the program requires other tasks to be resourced that are owned by program management and included in the approved program plan, and the safety tasks are not included in the plan. This builds in a conflict with safety because the need for safety resources will not be approved. Safety activities may be continuously delayed from one milestone to the next: the program moves from milestone to milestone without the safety tasks being executed, the safety organization lacks required resources, and the safety organization fails.

Executive Review

Since it is possible to launch an automotive product without properly performing all the system safety tasks on schedule, what motivation is there to include them in the program plan? These safety activities may be seen to cause a conflict for scarce resources, particularly in the early stages of the program while resources are being ramped up to specified levels. The training of the safety organization can also cause an ethical motivation: engineers and managers do not want to be responsible for producing a product that is not known to be safe. Safety training reinforces what activities and resources are needed to assure safety of the product. If the engineering process includes these activities, engineering can request the required resources and communicate this request through the proper channels so that the needed resources are deployed to the program.

Every organization must have a process to make these requests, in order to plan for the acquisition and deployment of resources to all programs. Still, how does this influence the behavior of the actors to execute these tasks when conflict arises, and when it is physically possible to ship a product without all the activities being executed? For example, suppose the safety expert needs to collaborate with the system engineers and hardware engineers to determine the safety mechanisms required to meet the hardware safety metrics. If other tasks for the system engineers and hardware engineers are given higher priority, these required safety mechanisms may be late or missing.

The priority of safety-related tasks is raised by elevating the visibility and importance of performing these tasks on time within the enterprise. Performing safety activities on time can be made a criterion for success for the company's executives so that the management of the program is empowered to allocate the necessary resources for safety in order to achieve a successful product launch that includes the safety tasks. Then the executives responsible for the programs will be motivated to support requests from program management concerning resourcing safety-related activities, because execution of these tasks supports the executives' personal success. Program management not planning for safety-related activities, not resourcing them, or not ensuring that they are completed on time will become unacceptable.

To ensure acceptable performance, program management can raise the priority of safety-related tasks when the program timing and status are reviewed. Trade-offs that delay or prevent the execution of safety-related tasks will not be allowed, and executive support for this position can be made visible.

This can this be accomplished even if failure to execute the safety tasks does not physically prevent product launch, by means of a periodic senior executive review. The visibility of this review is clear to the program team and management, so they will strive to accomplish a successful review. The review encompasses the execution of safety processes on every project. In a large organization, a review of this scope must be made as efficiently as possible: it should be at a high level and be clear and concise so that executive understanding is quickly achieved.

Such a review can be facilitated using safety metrics. Metrics have been successfully employed by quality organizations for many years, to show outgoing product quality, internal manufacturing quality, and the maturity of engineering organization–based standardized maturity models. Safety metrics measure the timely execution of safety tasks that have successfully passed safety assessment. The safety policy and safety process can establish when these tasks are to be completed and what artifacts are to be assessed as a result, and data can be captured from this process for every program. The criteria for passing or failing a due date must be made clear and non-debatable for this to be accomplished efficiently. Then these metrics are summarized for each program. By agglomerating these metrics to consistent program milestones, the status of the safety process can be measured for each project.

Having a favorable status can become a goal and motivation for each project team and program manager. Then project statuses are consolidated into overall metrics. This overall status can be organized by product and by each business unit or other organizational division of the enterprise. This allows senior executives to evaluate the performance of the enterprise with respect to safety and focus attention on those units needing improvement. Programs that do not meet the target status can be reviewed further by executives with the program owner and the functional owner, such as the lead executives for program management and engineering. Engineering executives and program management executives will become knowledgeable about the process and program status and recovery actions so that they can efficiently brief senior executives.

These periodic reviews need to be on a cadence of sufficient frequency for executive action to facilitate recovery of the program if necessary. Follow-up actions and a new status may also be included. Senior executive action is focused on programs that are not achieving the planned recovery. This review process provides an opportunity for senior management to be successful by executing safety tasks on time, thus mitigating potential resource conflicts. Divisional executives also share in this success by supporting safety-related activities and planning for this support to be available. If these safety tasks are not included in executive reviews, resourcing the tasks that are included may preclude enough allocation to the safety organization. The prioritization that the executive reviews provide may not be available for the safety activities, which may prevent execution of safety tasks – and the safety organization fails.

It is evident from this discussion that failure to meet any of the five criteria for success will almost certainly lead to a failure of the safety organization, or at least diminished success. In addition, meeting these criteria helps the VM or automotive supplier become more competitive. Developing or acquiring the required system safety talent will deny this talent to competitors of safety-related automotive systems. Having safety embedded in engineering also provides favorable optics when the organization is audited for a safety culture sufficient for the safety-related product. Depending on the extent of the career path for safety professionals, as compared to the career path in competitors' organizations, recruiting may be enhanced. The optics of seeing safety tasks tracked in the program schedule can increase trust and confidence that the VM or supplier is diligent with respect to safety and favorably affect business. The optics of the safety process being reviewed by executives can have a similar effect. In addition, when the metrics are sufficiently favorable, they can be used by an automotive supplier to hunt business. The VM's or automotive supplier's image improves, and the safety organization succeeds.

Pillars of a Safety Process

The system safety tasks required for the launch of an automotive product are identified (identification of the tasks for an individual automotive product application is one of the system safety tasks) and assigned to various functions to be accomplished during different phases of the product life cycle. There are tasks to be performed in the concept phase, such as the hazard and risk analysis; other tasks are performed in the design phase, such as hardware safety metrics. All the requirements need to be verified. There must be validation that safety goals are achieved and validation of the safety of the intended function. There must also be assurance that these tasks are performed correctly and on time. For example, if the hazard and risk assessment is not completed before the design is started, or if it is completed incorrectly, this may prevent the design from taking into account the achievement of safety goals. The hardware safety metrics may not consider all the safety goals or identify needed safety mechanisms.

The safety tasks and the assurance that they are performed correctly and on time form the *safety process*. All the safety tasks that may be required for safety-related process development are included in the safety process. The means to assess whether these tasks are performed correctly is also included in the safety process, as is the means to report their completion and periodically audit execution of the safety process itself. This safety process is intended to ensure that safety is considered diligently throughout the product life cycle.

The process creates evidence of this diligence, such as work products. *Work products* are artifacts that provide proof of compliance with process requirements. The work products may include the content of the work products included in the safety standards. The *safety case* is the compilation of this evidence that the safety requirements have been elicited and of compliance with these safety requirements, with the argument for diligence. The evidence is necessary for the safety case. With no evidence, there is no case.

This assurance requires audits of the process and assessment of the tasks. Process audits ensure that it is being followed properly and support resolution of any issues with a particular program in following the process. Assessing the tasks ensures that they are completed accurately and completely without omissions or errors that could affect the safety of the product or the safety case, so that there is evidence of diligence. The previous discussion concerned success factors required for an organization to accomplish this process. Each of these five success factors is critical to establishing a successful, robust safety process in the enterprise, and the organization must support achieving these five success factors.

The process requires some independence of groups of activities as its foundation. The system safety organization and its integration into the organization enable the creation of these independent groups of activities, referred to here as *pillars*. These pillars are discussed next.

A system safety process can have three pillars:

1) Determination of policy. This establishes the framework in which the system safety process is implemented. As discussed previously, it can be somewhat general, setting up requirements that each business unit of the enterprise must respect with its safety process. Or it can be more specific, especially if the enterprise is more restricted in its product offerings, and specify which functions are responsible for executing specified activities in support of the safety process. Regardless of the granularity of the process, the policy directs that a safety process be followed.
2) Audit and assessment of the safety process. These functions compose the second pillar and are distinct from each other. The assessments are evaluations of the process artifacts to independently determine their acceptability. When thousands of work products are being created in an enterprise, an unbiased evaluation of each work product to criteria established by the enterprise is needed, to maintain a uniform level of compliance with the system safety requirements. These criteria are determined by the audit and assessment functions. Then the audit function periodically reviews the execution of the process by each program.
3) Execution of the tasks in the safety process. In each program, the personnel associated with achieving the product launch are also responsible for the execution of safety-related activities. These personnel may include one or more safety experts to support the program team in this execution. This program team is independent of the audit and assessment personnel, as well as the safety policy personnel. The audit and assessment personnel also support program personnel with an independent perspective.

These pillars are usually expected to be independent. Establishing a policy, assessing compliance with the policy, and executing the policy are activities that should be performed without undue influence on one another. ISO 26262 [3] requires independence between the people executing the second pillar and the release authority of the product associated with the tasks performed by the third pillar. Different levels of independence are required, depending upon the rigor required to achieve the safety goals. Ethical implementation of the safety process requires that there is not even the appearance of influence of the safety assessment and audit personnel by the program team with deadline responsibilities for releasing the product.

It is not uncommon for the first pillar to be independent as well, although this independence is implemented as review and approval of a policy drafted by experts in the safety domain. This expert drafting team can include a cross section of safety experts from the business units within the enterprise. This helps ensure that all the elements needed in the policy are included, and that no requirements are misinterpreted within the context of any business unit of the enterprise. When the draft has been vetted by all stakeholders, it is submitted for approval. Some rework may be required after the review by the policy group; this is common for corporate policies.

Figure 2.1 reflects the relationships among these pillars. The policy that is approved for system safety is released by the enterprise policy group designated to approve such policies. This is the general corporate policy group or a specialized group with access to related domains, such as legal counsel, to provide adequate confidence for release.

The relationships illustrated in Figure 2.1 are not a reporting structure. The auditing and assessment function is not required to report to the policy-release function, and the personnel required to execute the policy and process are not required to report to the auditing and assessment function. Potential organizational structures, as well as potential advantages and disadvantages of alternative structures, will be discussed later in this chapter: these organizational structures are to implement the relationships shown in Figure 2.1.

The relationships shown in Figure 2.1 are intended to be *functional*. It is the function of the policy group to provide a safety policy for the auditing and assessment group to use. It is the function of the audit and assessment group to provide assessment and audit support to the program personnel responsible for executing the safety process and policy. The purpose of having these pillars is to achieve sufficient governance of the system safety activity within the VM or automotive supplier. These pillars enable the enterprise to systematically create a safety policy, enforce its deployment, and consistently apply the safety process in

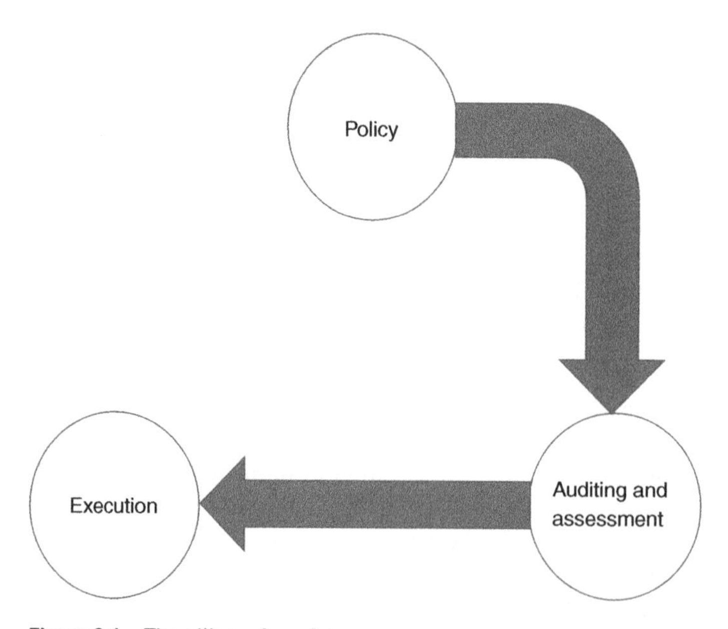

Figure 2.1 The pillars of a safety process.

compliance with the policy across the enterprise. This governance is to ensure sufficient diligence with respect to system safety as well as to ensure consistency in the practices used to achieve system safety in the products for release to the general public. Governance of the safety practices within the enterprise is critical in the automotive industry. Inconsistency can adversely affect the safety case and should be avoided.

Consider a large automotive safety supplier enterprise with a global footprint. Numerous manufacturing facilities are located to efficiently service VM customers with reduced inventory requirements and product-sequencing requirements for the VMs' manufacturing needs. This supplier enterprise also has multiple development centers in different countries supporting the same line of products for both local and global VM customers. The local VM customers have more dynamic needs to be accommodated, while global VMs have multiple requirements for versions of the product being developed for different global markets. Some local VM customers have minimal or diverse explicitly expressed system safety requirements. Requirements are fluid throughout the development and product launch cycle. Functionality is specified at the beginning and then changed later as the program unfolds, requiring updates to the concept and design safety activities performed.

Competition for local VM accounts is extremely price sensitive: sourcing may be won only with the most competitive bid. This can create a motivation to fulfill only those safety requirements explicitly expressed by the VM, since winning price concessions for changes or required safety analyses rework can be very challenging. Likewise, some global VMs have system safety requirements that are consistent, extensive, and explicitly expressed for every program awarded to an automotive supplier. Following these requirements is consistent with the supplier's internal processes. Compliance is monitored by the VM, and the supplier satisfies these requirements.

Chapter 1 discussed that the consumer expectation when purchasing a vehicle is that the risk of using the purchased vehicle is no greater than the risk of using vehicles already in the field. Regardless of the program, this consumer expectation applies. No concessions are granted to meeting this expectation based on an individual program. How can the supplier achieve this expectation in the previous example if the VM's explicitly expressed system safety requirements are not consistent among different VMs?

Each VM is also required to meet this consumer expectation, and the VM expects the supplier to support the VM in satisfying the expectation. The supplier cannot directly confront one VM with another VM's different system safety requirements, due to confidentiality considerations necessary to maintain trust and to continue to win business awards. Each VM has confidentiality requirements written into the contract for sourcing development support and series production, and sharing information with another VM violates the terms of this contract. So, another means must be found to resolve this issue.

The first pillar of system safety targets this dilemma. It is possible to achieve, based on the supplier's governance. The automotive supplier establishes a system safety policy internally, for which compliance is required on every program. The relationships shown in Figure 2.1 ensure that this policy is applied consistently: they are established for this purpose. Evidence of this compliance produces a consistent safety case. The system safety process provides artifacts that provide this evidence, and these artifacts are assessed to provide consistency. Since it is the supplier's policy, and not the confidential information of any VM, the information can be shared with each VM. The supplier's artifacts may not exactly match the artifacts required by a particular VM; or content may overlap between the VM's artifacts and the supplier's artifacts, although in total they may align. Exceptions are argued on an individual basis to meet additional requirements a VM requires. Different formats for the artifacts or grouping of the evidence may be requested; and different types of analyses, such as quantitative analysis versus qualitative analysis, need to be resolved.

The automotive supplier can use efficiency of scale to be competitive when its system safety policy requirements exceed the VM's explicitly expressed system safety requirements. How can efficiency of scale be accomplished if there is only one VM or only a few VMs without these system safety requirements? The supplier builds compliance into its basic platform or the platform that is being reused or modified for highly price-sensitive bids. It may be more efficient to include compliance with the safety requirements already elicited by following the supplier's process consistently than to modify the product. The analyses can be reused by considering the impact of differences between the available product platform and the particular version specified by the VM. Updates to the policy can ensure that the system safety requirements meet consumer expectations. These updates can be drafted by safety personnel and approved by the policy group. This provides further evidence of diligence. The safety case is consistent, and the policy is its pillar.

As with any policy, interpretation of the policy requirements is always necessary. Many different safety experts across the enterprise will read and form an interpretation that is influenced by the context of their safety function. The policy requirements are broad; interpretation is required so compliance can be planned and achieved for each program. All the required activities must be planned, resourced, and executed for each program. This interpretation must be consistent across the enterprise, in order for the policy to be consistently followed and to have a consistent safety case.

Consider the automotive supplier in the previous discussion. This is a global tier 1 supplier with many widely dispersed facilities that all contribute to achieving the safety requirements of the automotive products supplied. Different technical centers in different countries service multiple customers with multiple products. There are different local languages, different local cultures, and different local management organizations and management styles. The safety demands of the customers are different: they have different requirements, or minimal requirements for system safety.

The development teams are organized into teams to focus on specific customers with minimal communication required among the teams. The primary focus of each team concerns their customer. They interpret the safety policy with a nuanced understanding based on interaction with the customer they service.

Further, the technical centers are locally focused with minimal communication required between technical centers. Programs with their customers do not require extensive interaction with personnel at other technical centers after the program begins. Tailoring of global platforms is executed locally to provide more efficiency and responsiveness to their customer. Leadership at the technical centers demands that this local focus be maintained. It is a measurable goal of local management to provide this responsiveness and acquire follow-on business and additional wins. How can consistent policy interpretation be ensured? How can compliance be consistent?

While the diversity just described supports disciplined product focus, disciplined policy interpretation also requires focus. Reports to senior executives, such as safety metrics, must be based on a consistent interpretation of policy in order for these executives to accurately steer the enterprise toward its safety objectives. A clear understanding of what constitutes compliance with policy must be conveyed to every program team of the enterprise. To do this, the language of the policy needs to be culturally neutral and consistent with the terms used in each business unit.

This can be achieved by a group of safety experts committed to achieving a consistent policy interpretation among them. There should be representation from every business unit of the enterprise. They share experiences and agree on interpretation of the safety policy in each of these shared experiences. These experts have support from their business unit for review and discussion in order to vet the policy for consistent interpretation.

This group of safety experts is deployed to the many different technical centers. Their role in each technical center ideally includes technical safety leadership so they can help form the consistent interpretation needed. However, they are centrally led, independent of the release authorities at the technical centers, with a focus on achieving consistency.

Central leadership helps maintain a consistent interpretation of the policy among the safety experts. If a conflict arises among the experts with the interpretation of the enterprise safety policy, this leader resolves it. To cement understanding concerning interpretation as well as practices to help achieve the objectives of the policy, this leader publishes guidance with respect to the policy interpretation. The expert team supports and endorses this guidance in order to facilitate its deployment.

With this consistent policy focus in mind, these experts independently assess the system safety tasks and work products for each program and for the basic platforms used. This independence satisfies the requirements of ISO 26262 [3] for independent assessment, as well as ensures the consistency of interpretation and therefore the accuracy of the safety metrics for the enterprise. Misunderstanding of any safety requirement by any individual on any program team can be independently and consistently resolved across the enterprise.

Consistence guidance documents can be used, such as checklists, which are provided centrally and available not only to the assessors but also to the authors of the safety artifacts. Use of the same assessment guidance by the assessor and the author helps with more efficient assessment and better safety metrics for the program.

This central group also audits the execution of system safety tasks within the process used to achieve compliance with system safety policy. The auditor may audit programs from multiple product lines or business units; this prepares the auditor not only to provide a consistent judgment of the program status, but also to spread best practices for resolution of issues and recovery from missed milestones, if possible. If a consistent audit method is employed, each program will have accurate expectations of what to expect during the audit and can plan to facilitate making the audit as efficient as possible. This can economize the time spent in the audit and away from other program tasks. In so doing, the resulting documentation can be compiled into metrics that are used for executive review. These audit reports can relate the program status with respect to the relevant milestones. Issues impacting the program team can be addressed as well as other findings. The findings in the reports will help prepare program and executive management for senior executive reviews, thus ensuring consistent interpretation of safety policy. Assessment is a pillar.

Ensuring that consumer expectations of safety are always met requires more than a policy and independent assessment experts. Having a consistently interpreted and meaningful policy, as well as independent safety experts to review and report compliance, is necessary but not sufficient. Significant effort is expended in development to perform safety analyses, elicit system safety requirements, and comply with those requirements in order to launch products with the level of safety in the field today. Each safety-related system that has been launched since 2011 considered functional safety and perhaps the safety of the intended function. Analyses were performed at every stage of the product life cycle.

ISO 26262:2011 has 130 normative work products for functional safety, and VMs and tier 1 suppliers aspired to comply with requirements addressed by these work products even before the standard was released. Many considered this to have become the state of art after its release.

Even more effort is required for the level of system safety in the field today. Many test-track and field trials have been performed to properly develop advanced driver assistance systems. Reference [9] provides some guidance for validating aspects of these systems and was considered prior to the launch of some systems in the field. Parts of these practices are being standardized in ISO PAS 21448 [4], with many years of global participation. Participants offer the practices they already employ as candidates for the standardized methods. Therefore, many of the methods and measures that are being standardized in [4] are already deployed in launched vehicles and are reasonably expected to have been deployed by consumers for future launches, regardless of whether the standard has been released.

All of this implies that a substantial effort is needed to meet consumer expectations for products not yet launched, to execute system safety tasks at least equal to those executed for automotive products already in the field. Consumers will not tolerate assuming additional risk when they purchase a vehicle because an enterprise that produces safety-related automotive products did not plan the tasks, deploy the resources, or manage the program to get the expected activities completed by the required milestones. While it is perhaps not a conscious expectation, consumers do not expect to assume risk due to a manufacturer not successfully executing tasks required for the safety process. These system safety tasks involve activities by safety experts as well as others involved in the development and industrialization of automotive products. System analysis and specifications are required from system engineers, software safety analysis is required by software engineers, and hardware safety analyses are required by hardware engineers. Program management must oversee the tasks required for the safety process, perhaps assisted by a safety manager. This needs to occur systematically for diligence. The entire enterprise is involved.

Consider how safety is realized in the sequence of tasks from initiation to termination of an automotive product. The automotive product initiation starts prior to a request for quotation (RfQ) when the market features are being determined. These market features lead to procurement of systems that can provide these features. During the preparation of the RfQ, the VM determines the safety considerations needed from the supplier, to ensure that consumer expectations are met by the product supplied. These include expectations of the function itself and its acceptable limitations, as well as the safety consequence if the system fails in order not to put the consumer or general public at an unacceptable level of risk. This requires some early safety analyses by the VM of how the product is integrated into the vehicle safety architecture, possibly including a HAZOP type of analysis of the system functions at the interfaces with the other systems in the vehicle architecture. Depending on how the function is integrated, the possibility and consequences of foreseeable abuse may be considered. All of this is to ensure a level of diligence consistent with what the consumer reasonably expects, prior to initiating the development and deployment of a safety-related automotive system.

In preparing the quotation, the supplier must determine the level of resources necessary to ensure that the supplier's system safety policy is implemented, assessed, and audited, as well as any updates of the safety case produced by this policy that are required to meet additions specified in the RfQ. A similar program can be used as a starting template and resources modified to accommodate the unique requirements of this program. Also, any outstanding conditions of acceptance from the similar program need to be planned and resourced so they do not remain outstanding and prevent acceptance of the program under consideration. This diligence supports executing the safety process and helps meet consumer expectations.

It is assumed that the VM will require a safety culture at the supplier sufficient to meet consumer expectations of the supplied product. The existence of a comprehensive and unambiguous safety policy is a starting point for this safety culture. The priority put on system safety by the executives of the enterprise is expected to support a strong safety culture. For this to be effective and these assumptions confirmed, there must be a method of communicating the detailed actions required by each individual in the organization to ensure the safety of the products launched. This requires the enterprise to have evidence that the development engineers, managers, and executives have been trained in system safety. This training needs to provide not only a general overview of system safety, but also the actions required for each of their roles to successfully implement the safety process for the enterprise. Often this requires multiple specific trainings tailored to each of the disciplines involved, at appropriate levels of granularity for both senior executive and software engineers. Then these engineers and managers execute system safety tasks as required by the system safety process derived from system safety policy.

Even after the training, engineers and managers need expert support to enable them to accomplish their tasks on time and accurately in the program. Some analyses are best performed by a team that is facilitated by a safety expert familiar with the analyses. An allocation for the required safety experts can be included in the quotation for this role as well as to interface with the VM's safety expert and to facilitate execution of the system safety process. Knowledge of the expected complexity of this interface, and the needs of the customer interfacing, should be considered when determining the allocation for a particular program; some customers require resource-intensive interface practices.

System engineering and the other engineering disciplines work together with verification personnel to ensure that there is evidence of compliance with the safety requirements of the system and subsystems. This includes constructing the requirements so that all needed attributes are included, and that the verification method is linked. These verification methods must be very specific, such as detailed description of the test case with the pass/fail criteria. The results of the verification must also be linked.

A similar activity is carried out in the preparation of the production process. The analysis of the design and potential failures must be consistent with the analysis of the production process and potential failures. Even though these analyses are usually performed by different people, inconsistencies – such as differences in severities in the design FMEA and the process FMEA – can adversely affect the safety case. Then production is monitored, as are field returns for system safety anomalies. The production quality process often provides this monitoring of production and field issues. This process normally includes a method for resolution, permanent correction, and closure. Then safety is maintained throughout decommissioning. Consideration of the decommissioning personnel is paramount: often, specific instructions and warnings are warranted. All of this requires system safety execution. Execution is a safety pillar.

From the previous discussions, it is evident why policy, auditing and assessment, and execution are considered the three pillars of a system safety organization. The policy process is fundamental to a uniform deployment of a system safety process. It provides the legitimacy needed to authorize deployment of resources to implement the process. Auditing and assessment are needed not only for governance, but also to ensure that the process is consistently implemented. Without execution, the process cannot be implemented, and this execution consumes the most resources. The three pillars are independent from each other, and they are used to build the system safety organization. The organization must provide the functions required and must provide the independence required.

What are the alternatives for such an organization? It can be deployed throughout the enterprise in numerous ways, since each enterprise already has an organization for developing and launching automotive products for production. The interfaces and management considerations will be different. What are the advantages and disadvantages involved in the trade-off to determine the organizational structure? Are the critical success factors a consideration? Is integration affected by the organization? What about personnel development? These questions will be addressed in the next section.

Alternatives, Advantages, and Disadvantages

In the previous discussions, the activities of the safety organization and the critical success factors were listed. The importance of each of the critical success factors was explained, and the activities involved were put into the context of launching an automotive product. These need to be taken into account when organizing or reorganizing a system safety organization. The shape and deployment details of the organization can enhance or deter the achievement of the critical success factors. The organization may place obstacles in the way of success or help ensure a successful safety process. All the listed activities need to be resourced and organized so that they support the enterprise's needs and are performed consistently. How they are organized, and how the resources are managed, can affect the efficiency and quality of execution of the system safety process.

Determining the content of the safety policy needs to include individuals familiar with all three pillars of system safety. The enterprise policy organization can contribute to the consistency of the safety policy with other policies as well as the mechanics of releasing a corporate policy. The auditors and assessors provide technical knowledge required for the content as well as the mechanics of performing the assessments and audits, critical considerations for consistency that need to be specified, and efficient reporting and data reduction. The safety managers involved with interfacing with the customer and supporting the product development and operations personnel can also advise about critical considerations concerning execution. Likewise, developing and maintaining a safety process must include consideration of the development process the enterprise has determined is applicable at a high level for all products developed by the VM or automotive supplier.

This process includes milestones in every program that are consistently named and understood. Such milestones trigger management or executive review on a periodic basis. Existing summary reports by business unit can be used in the management of the enterprise. There may even be consistent formats that foster easier understanding by the reviewers. The safety process must be compatible and achievable with these product developments in order to execute systematic analyses of the hardware and software systems. Adapting the safety process to the existing product development process allows consistent review, auditing, and assessment, so that corrective action or recovery can be implemented before process violations occur. This feedback supports management of the resources associated with executing the safety process.

In addition, the organization must provide the necessary system safety training for all enterprise functions involved in launching and servicing a product. Deployment of basic safety training helps foster a safety culture and can be done uniformly across the enterprise, perhaps even remotely or online. Specific domain training requires more human interaction and can be performed by the assessors who cover the technical centers involved. The training is extended into the daily tasks on the job and involves coaching and development of safety experts as well as development personnel involved in other engineering disciplines. Development of safety experts can take years and requires close communication as well as a personal mentor who can serve as an example to develop the calibrated safety judgements required day to day.

Independent auditing and assessment of work products is required, as well as reporting, and must be supported by the organizational structure. The organization must be clear about this independence when reviewed by external parties. The organizational structure should provide clarity about who is involved in each activity for every automotive product provided by the enterprise. Any ambiguity in responsibilities can be a potential source of error, duplication, or omission. While such issues may be resolved, significant efficiency is lost. This inefficiency due to the ambiguity of responsibilities is repeated until the organization is clarified. The organizational structure should support efficiency, and various alternatives can achieve this.

Consider a global VM or automotive supplier with multiple safety-related product lines, some or all of which have both electrical and mechanical content. Each product line includes engineering organizations that include system, software, and electronic hardware engineering groups. These groups may be deployed in one technical center or among multiple technical centers. The product lines may serve global customers. Each engineering group is responsible for part of the development of safety-related products: hardware engineering has standard platforms that are deployed and adapted to its customers and also develops new platforms to be reused, and the software group pursues a similar strategy. Each engineering group is managed by someone with deep domain knowledge for that engineering discipline as applicable to the product line; this person has advanced within the group and is familiar with all the technical aspects of the tasks performed, as a result of personal experience. The engineering groups have the focus and depth of knowledge needed to design the automotive products. The leaders are able to mentor the engineers and continue to advance the technology. Each engineering organization incudes an executive engineering management structure, which allows clear direction to be maintained. Issues and priorities can be resolved consistently, because engineering reports to its product line. This is a sustainable structure.

To include system safety in this structure, one alternative is to add a system safety group to engineering in each product line. This system safety group can be structured like hardware, systems engineering, or software engineering, with reporting into the product line leadership mirroring the reporting of the other engineering domains. Each product line acquires a system safety domain expert to manage the system safety group: this expert may be a transfer from the central system safety auditing and assessment group or can come from outside the enterprise. The safety domain expert has or acquires domain knowledge for the product line assigned: this product line knowledge may come from reviews, if the expert is transferred from the central safety auditing and assessing group; from a competitor, if the expert comes from outside the enterprise; or from previous consulting. Otherwise, product knowledge must be acquired on the job. In addition, this expert often has domain expertise in system, software, or hardware engineering, or a combination of these domains. This knowledge is useful for understanding both the technology and the safety analyses that are required. This understanding can aid the expert in better leading the system safety group toward its mission.

The mission of the system safety group is to support the programs of the product line with safety analyses and other system safety tasks, as well as to facilitate execution of the system tasks for which the other engineering domains are responsible. For example, a member of this group can serve as the safety manager, as referenced in ISO 26262. A member can perform the hazard and risk assessment. Determining hardware metrics to achieve required numerical targets involves considering the automotive safety integrity level (ASIL) of the hazards and safety goals. Software safety analysis by the software engineers may involve mentoring and coaching by a member of the system safety group.

Also, a member of the system safety group can help determine the criteria for acceptance for the three levels of validation of a level 3 driver assistance system, as described in ISO PAS 21448 (safety of the intended function [SOTIF]). First-level testing identifies triggering events. When those have been determined, the second level of testing on a test track evaluates triggering events supplemented by simulation. Then systematic hunting for additional unknown scenarios resulting in triggering events occurs in the public domain with further simulation.

The system safety group can tailor a process to support the product line engineering process and to meet the safety process of the enterprise. Work product content, such as safety analyses, is added to existing work products established by engineering quality in order to provide artifacts to be used as evidence in the system safety process.

The staffing of the system safety group is determined by the product line programs' demand for support. System safety staff budgeting is established by agglomerating the allocations of each committed or target program for anticipated system safety support. Staff can be mentored in the group by the system safety group manager as well as senior members of the group. Additional mentoring results from interactions with the central system safety assessment and auditing functions, as well as the continuous system safety training provided across the enterprise. This promotes consistency in the group, and as a result, execution is consistent.

It is clear that independent auditing and assessment are needed to support this group. The group is deeply embedded in the product line organization that has release authority for the product. Therefore, the group does not provide the independence needed for assessment of work products and auditing of the process as required by ISO 26262 [3]. To ensure consistency of interpretation of the enterprise safety policy when performing audits and assessments, a central system safety group can provide the auditing and assessment instead of an external safety expert engaged by each product line. Different external assessors may not necessarily agree on all criteria or be aware of the criteria established by the enterprise policy and process.

This use of a central safety auditing and assessment group is supported in ISO 26262 and can potentially provide enhanced evidence of independence if the cost of the audits and assessments is not directly charged to the product lines being assessed and audited. If fluctuations in demand require external support, the central group can engage external safety experts and monitor them to ensure consistency. The policy, guidance for policy interpretation, criteria, and support – such as checklists and a "back office" (e.g. specialized support from low-cost countries such as India) to review documents to be assessed consistently using these checklists – may be provided to external safety experts if engaged by the central group.

The central system safety group is directly funded by the enterprise, in order to further enhance the optics of independence as discussed. Budgeting for this central group can be accomplished independently from the product lines by directly reviewing the marketing forecast and projecting resource needs. This will provision for satisfying anticipated auditing and assessment requirements without dependence on the product groups that are audited or assessed. The central group can develop a central process for system safety that satisfies the enterprise safety policy. This process is vetted by the various enterprise product groups to ensure acceptance; each product can then use the central process directly or tailor a process for its product group that is compliant with the central process but crafted to respect the needs of the product group and its structure. Product line processes used for each program can be audited to ensure that any tailoring by the product line is in compliance with the enterprise safety policy. If issues arise during a program audit, resolution is addressed by audit actions for that program. If the issues are more generic – for example, with the tailoring process for the product line – then they can be escalated if necessary, to be addressed by the safety group of that product line.

System safety tasks and work products are assessed by the central auditing and assessment group as required, to ensure compliance with the enterprise system safety policy. This central assessment of work products helps maintain an appropriate, uniform level of rigor across the enterprise for products with similar safety criticality. This evidence of consistent rigor supports the evidence of diligence for the respective safety cases. The central auditing and assessment group must also comply with the enterprise system safety process. The process will specify the requirements for assessment and auditing, including, for example, the frequency of audits and the topics to be addressed during the various life-cycle phases of each program audited. There is also guidance for how the audit is to be conducted and the roles of the required attendees.

The system safety expert who manages the central safety organization may also serve as an auditor. In that role, the safety expert has direct visibility of compliance with the central process for auditing as well as assessment. Using this process visibility, the managing safety expert can ensure compliance of the central auditing and assessment group with the enterprise safety policy through the central process, and supplement with the use of tools such as system safety process software and checklists. The central process software can be customized to tabulate the execution and assessment status of each program using standard attributes and also agglomerate these data for analysis. This analysis can determine systemic issues in the execution or assessment of the process, which can then be addressed. External review can supplement this assurance and can be especially useful when the central process is being established. This also ensures that the central process respects any standards envisioned to be satisfied by the enterprise system safety process. The goal is to obtain compliance with these standards by constructing the central process. Then execution of this process can achieve compliance with these standards.

The central system safety organization can deploy consistent training for product line executives, managers, and engineers to support this compliance. This training also supports the system safety group in each product line by providing training for personnel in the domains the safety groups are supporting. Training product line management supports the validity of the mission of the product line safety groups. Evidence of this training helps demonstrate a safety culture to external process audits: training records can be examined, as well as the rate of participation, and the content of the training can also be made available. Such training also improves the efficiency of the product line system safety process execution. Timely execution of safety analyses is supported by training that teaches exactly how to perform the needed analysis. This promotes accuracy and promptness of analysis at the correct program phase. If completing such analysis is included in performance reviews, successful execution is directly rewarded.

Training the engineering groups reduces the time required for facilitation in each program. Personnel have the necessary motivation and skills and require only limited support and mentoring by the product line safety group. Leaner system safety deployment is supported, and program efficiency improves.

This enterprise organizational relationship is illustrated in Figure 2.2 as alternative 1. The illustration is simplified to show identical structures in each product line for clarity so

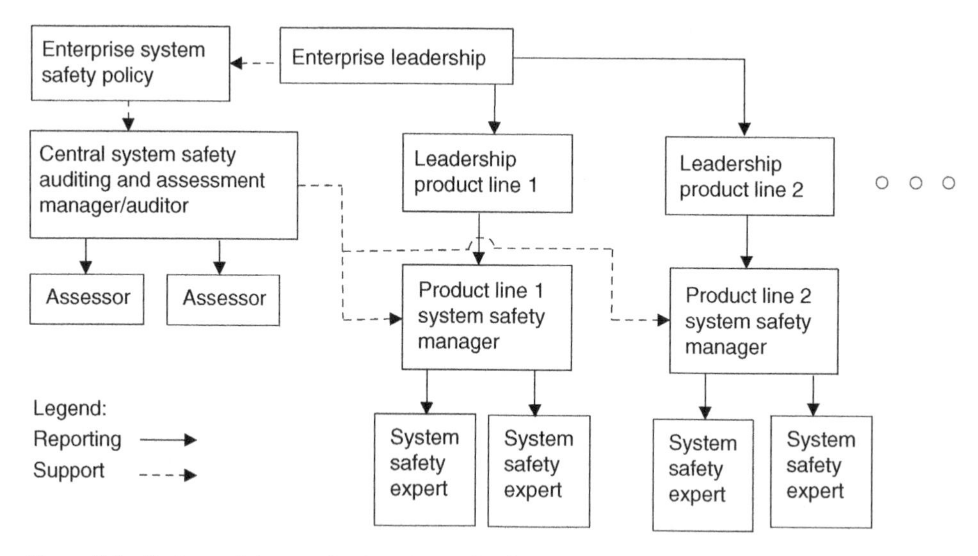

Figure 2.2 System safety organization alternative 1.

that the functional relationships are consistent. The system safety policy group for the enterprise is supported by enterprise leadership; this is critical for the policy to be respected as required and expected by senior executives. This support can be in the form of a system safety policy board or group, or as part of a central enterprise policy administration that includes other policies approved by enterprise leadership. It is possible to have both: there may be a safety policy board for review by several enterprise areas, such as legal counsel, quality leadership, and engineering. Then, in support of this board, the enterprise policy administration may manage the release and distribution of the policy. This enterprise system safety policy then supports the auditing and assess for compliance with this policy. Execution is managed within each product line: leadership for each product line is supported by the product line safety group, and each product line safety group is supported with auditing and assessment. Auditing and assessment are centralized.

A second alternative is illustrated in Figure 2.3. There are many similarities with the first alternative, such as enterprise leadership supporting the system safety policy, and the system safety policy supporting the central system safety auditing and assessment group. Here, the enterprise is a global VM or automotive supplier with multiple product lines, and each product line has its own dedicated engineering groups that are managed separately from the other product groups. The engineering groups include, for example, systems engineering, hardware engineering, and software engineering. For simplicity, only one of these engineering groups is shown in each product line in Figure 2.3.

In this alternative, rather than have a dedicated system safety expert managing the system safety group, safety experts report to an appropriate manager. This manager may or may not have experience as a safety expert, but needs safety support in order to fulfill the requirements of the enterprise safety process or the safety requirements of a customer. In practice, these experts are consultants or contract engineers with safety expertise. They are hired into the engineering group due to their expertise in that discipline but have background as a safety expert or have had special training. In Figure 2.3, the engineering groups to which

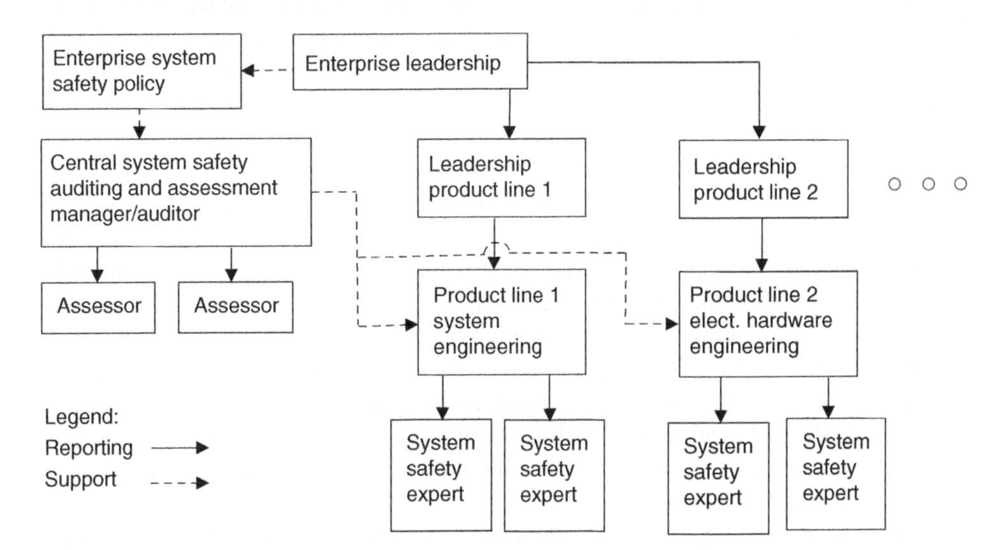

Figure 2.3 System safety organization alternative 2.

these safety experts are assigned are managers of a systems group for product line 1 and an electrical hardware engineering group in product line 2. Alternatively, safety experts can be assigned to more than one engineering group within the same product line. The engineering managers of these groups can mentor the safety experts in the product line and provide them access to product line documents and personnel. The engineering managers can also facilitate and support the safety experts managing compliance with the safety process, organizing the execution of safety analyses, and completing artifacts for safety case evidence. The engineering group managers are not expected to mentor the safety experts in the domain of safety; in practice, the safety experts provide detailed insight of the system safety process to their managers. Safety mentoring of the safety experts should be accomplished by the central safety organization – or it may not even be necessary.

Meeting the requirements of the central safety process and customer safety requirements becomes the responsibility of the safety experts. If a tailored product line process compatible with the enterprise system safety policy has not been developed, and the product line leadership and management of each engineering group do not think developing such a tailored safety process falls within the domain of their group, safety experts are deployed to support the programs in the product line. The process and tools available may be considered sufficient for this purpose: program demand drives staffing.

The enterprise relies on the central system safety auditing and assessment organization for mentoring and training. For example, in the systems engineering group of product line 1 in Figure 2.3, the system engineering manager mentors the other system engineers in the group. However, this manager is not expected to mentor the system safety expert: this mentoring task is left to the central auditing and assessment group. It is accomplished during the review of artifacts as well as in association with scheduled audits. These activities also provide support and credibility to the safety experts when resolving findings from safety analyses.

Consistency is organizationally supported by the system safety organization with no solid line reporting. The acceptance criteria for safety assessments is centrally enforced and reported. Findings from audits are also centrally reported to and resolved within the product line, to ensure consistency. Thus, the role of the central system safety auditing and assessment group is like alternative 1: independence is still provided, and training is deployed centrally.

Figure 2.4 illustrates a third system safety organization alternative. The enterprise is the same as in alternatives 1 and 2. Enterprise leadership supports the system safety policy, and the system safety policy supports the central system safety auditing and assessment group. In this alternative, the system safety manager for each product line reports directly to the central system safety organization. This way, the solid line manager is focused completely on system safety and is outside the product line hierarchy.

The system safety manager can report directly to an assessor assigned to that product line or through another hierarchical manager, perhaps in a matrix arrangement with the assessors. In either case, the product line system safety manager is managed by an experienced safety expert. Mentoring of the product line safety manager is by the central system safety assessment and auditing organization, as in alternative 1, but now it is augmented by the reward system of direct reporting.

The system safety manager provides the same support to the product line as in alternative 1. Support levels are maintained to encompass all the safety-related programs of the

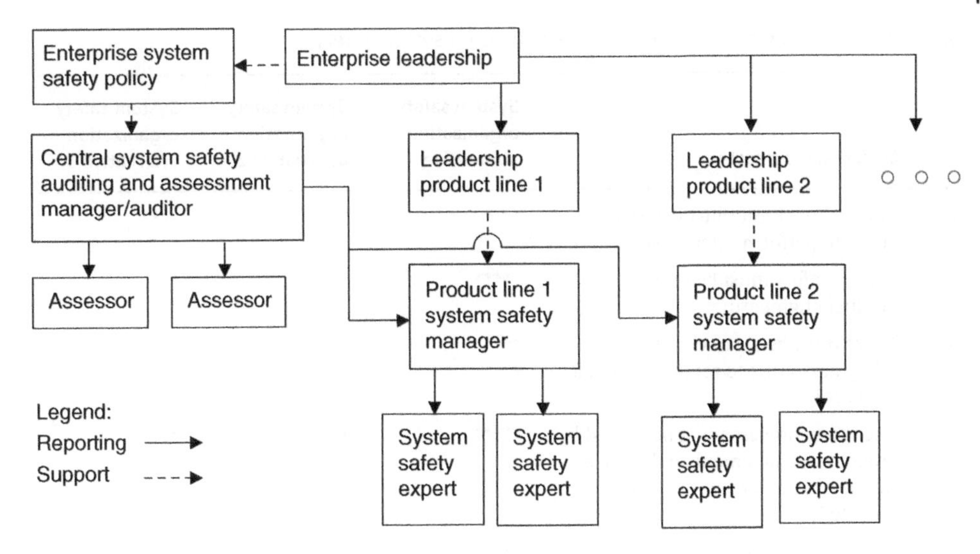

Figure 2.4 System safety organization alternative 3.

product line. Product line leadership provides the support necessary to access needed program documentation and personnel. This supports product line engineers who perform the needed analyses in order meet the system safety milestones for each program.

Consistency among product lines is ensured by direct reporting to the central organization. Mentoring is also provided by the product line safety manager to product line personnel, with the support of the central system safety organization. Specialized training can be organized as required and can be customized for the various engineering domains and management levels. Auditing and assessment are independent of the product line and provided by the central auditing and assessment organization, which fulfills the independent assessment requirements of ISO 26262. This independence is not compromised by direct reporting of the product line safety manager: this provides independence beyond what is required, and system safety facilitation is provided.

What are the advantages and disadvantages of each of these three system safety organizations? The discussions of each alternative hinted at some advantages and disadvantages, and others may be perceived, depending upon the vantage point in the enterprise organization. For example, a system engineering manager may consider that not being expected to mentor the safety expert is an advantage of alternative 2. Advantages and disadvantages can also be evaluated with respect to the critical success factors. These can be juxtaposed with each other for a systematic comparison, as shown in Table 2.1.

The extent to which each alternative ensures that necessary talent is available for each program of the product line, to comply with the enterprise system safety process, is highest for alternative 3. This alternative has a larger pool of talented system safety experts that may be made available to support all enterprise product lines, because this combined pool of talent is maintained centrally.

Alternatives 1 and 3 each have an expert safety manager dedicated to ensuring that the necessary talent exists based on projected workload, and capable of mentoring system safety experts continuously on the enterprise system safety policy. This function of the

Table 2.1 Evaluation of organizational alternatives for system safety.

#	Critical success factor	System safety organization alternative 1	System safety organization alternative 2	System safety organization alternative 3
1	The organization must have the talent to perform safety tasks.	****	***	*****
2	System safety must be integral to product engineering.	*****	*****	****
3	There must be a career path for safety personnel to retain them in a safety role.	***	**	*****
4	The safety process must be owned by program management so that tasks can be planned, resourced, and executed.	*****	*****	*****
5	There needs to be a periodic executive review cadence to ensure that the process is followed.	*****	*****	*****

product line safety manager is not altered by the difference in the reporting structure for the product line safety manager. The product line safety manager maintains the team of safety experts for the product line to support the engineering domain personnel accomplishing safety-related tasks for the programs on time. However, in alternative 3, the central pool of talented safety experts can be rotated or temporarily reassigned from one product line to another to cope with peaks and valleys in the workload. This can provide some efficiency overall in the execution of certain tasks. For example, an expert in executing hardware safety metrics can be deployed to programs where this is the next critical task to maintain on-time performance. System safety experts with significant experience in software safety analysis can be deployed to a program experiencing difficulty in that activity. This approach also adds to the system safety skill set of the system safety experts who are being rotated: they are rotated to a program and stay assigned to it for its duration, and become knowledgeable on the content of that product and the variety of tasks needed to complete its safety case.

The manager of the safety experts in the product line in alternative 2 is not a system safety expert and may not have performed or witnessed the execution of the safety activities of each phase of the safety life cycle. Enterprise leadership supports the system safety policy, and the system safety policy supports the central system safety auditing and assessment group. The level of effort to perform the various safety analyses based on complexity and scope may not be in the manager's background. Advance estimates of the level of resources needed are not based on personal experience, but rather are based on analogous activities within the manager's domain, and thus may be less accurate and effective. Alternative 2 was de-rated.

Concerning a career path in safety, the three alternatives diverge. The organizational structures are different, so the career advancement among organizational lines is also different. All three alternatives can provide technical ladder advancement in the safety role. In every case, each safety expert is provided some degree of individual mentoring from the central safety auditing and assessment organization. Centralized formal group training is also provided. As the skills of the safety experts are honed and proficiency is demonstrated, the organization may provide opportunities for advancement based the value added of the improved system safety skill. A safety expert can gain recognition both inside and outside the enterprise for competence and judgment.

With respect to increased management responsibility while remaining in a system safety role, alternative 1 provides an opportunity for safety experts to advance to product line safety manager or be promoted into the central safety organization. These are the two logical paths because of the solid line reporting, as well as similar opportunities in other product lines and support from the auditing and assessment organization. The latter is limited, as the central organization consists of auditors and assessors. Competitive candidates include experienced safety auditors and safety assessors. Nevertheless, internal candidates have an advantage of product and safety process knowledge as well as performance that has been observed by the hiring manager of the safety audit and assessment organization. Alternative 2 is further limited because advancement to product line safety manager is removed. There are no system safety management positions in the product line organizations for alternative 2. The potential for advancement in the system safety audit and assessment central organization exists for safety experts in alternative 2, and this may be the only possibility to advance organizationally in system safety. Alternative 3 provides the greatest opportunities for career advancement in the safety organization for both auditors and assessors as well as product line safety experts and managers. It provides the largest organization with the greatest need for system safety line management, and also may provide matrix management positions. Technical advancement opportunities include advancement based on demonstrated proficiency. Also, advancement to non-management assessment positions is available. This makes the enterprise more competitive in the market for acquiring safety talent: with a strong organization of professional safety experts, the organization is very attractive to other ambitious professional safety experts. Opportunities for system safety careers are supported and are more plentiful than in enterprises that are organized differently, which provides motivation and rewards for improvement. Alternative 3 is clearly best here.

Ownership of the safety process by program management is not significantly affected by the safety organizational choice. Whether the safety experts supporting the programs come from an organization like alternative 1, 2, or 3 may not significantly influence the planning, resourcing, and program management of system safety tasks. The central system safety auditing and assessment organization is not different, with respect to the auditing or assessment functions, from one alternative to another. This organization compiles metrics on compliance based on data compiled from the assessments of safety process artifacts and auditing; the data include timing, including lateness, as well as other numerical metrics for insight into the performance of the programs with respect to the system safety process. Then the immediate and root causes of non-compliances are determined, including lack of proper planning and resourcing as well as communication issues within the organization and its safety experts. Lack of ownership of safety tasks by program management can be identified if the immediate cause is that the task was not resourced and executed on time.

This is the motivation for having program mangers own the safety tasks. Program managers are not expected to perform these tasks personally – they are expected to get them done. This is expected for all three alternatives, and no distinction is noted.

Fulfilling the need for a periodic executive review is also not significantly affected by the safety organizational choice. The enterprise executives needed for reviews are the same in each organization. Support of the policy by enterprise leadership is most important: without it, support of the system safety process by the enterprise is difficult to achieve. Leadership support empowers the enterprise to strictly follow the system safety process.

An enterprise may consider having the system safety organization as part of the quality organization with the intent that then the quality metrics will include safety. Quality organizations in automotive enterprises have been providing meaningful metrics for decades. Engineering quality process models have structured maturity models that enable engineering quality metrics to be presented in a uniform manner with standardized criteria. For this reason, including system safety metrics may seem logical. However, there are potential pitfalls in this approach. System safety and engineering quality are not the same. Sampling projects to determine process maturity is not the same as rigorous examination of every analysis and artifact of every product launched for details that adversely affect the safety of the product or the completeness of the safety case. While the quality organization is expected to be skilled in metric presentations, it is not expected to be skilled in system safety. System safety is almost exclusively an engineering domain, very similar in rigor to other design domains. Determining whether the system safety requirements are complete as well as whether compliance with them has been achieved is be expected to be in the domain of quality. During the presentation of enterprise data to senior executives, system safety metrics are not presented, or are presented after all the quality metrics are discussed and only if there is time available. While quality executives provide valuable review and insight into system safety metrics, they are not the preferred personnel to present and explain those metrics.

Further, system safety experts cannot be mentored in system safety by quality management. They may gain skills in the domain of quality but not in the domain of system safety. Typically, system safety experts are skilled engineers who have also gained expertise in system safety. They may be system engineers or software engineers with detailed domain knowledge in safety-related system design who have also acquired proficiency in the domain of system safety. This facilitates their ability to support other design engineers in their system safety activities. For this reason, they command a higher price in the labor market than quality engineers do; this is likely to continue as the demand for system safety experts increases due to the growing number and complexity of driver assistance and automated driving products. Thus, salary equity with quality personnel is problematic if system safety personnel are in the quality organization. An approach other than combining quality and system safety is wise. Still, the functions of engineering quality benefit the execution of the safety process: quality must support system safety.

Communication channels for information that affects system safety can be affected by the organizational choice. Where the boundaries are placed within an organization determines where communication across boundaries is necessary. Consider communication of design changes during a product's development that could affect safety in an automotive program. These changes may be electrical, such as changes to the microprocessor or the software operating system. In alternative 1, the product line system safety manager watches for such changes and can set up a product line process that includes providing alerts to the system safety manager when such changes are contemplated. The product safety manager has direct responsibility for the safety experts and is supported by product leadership to implement such a process.

In alternative 2, there is no product line system safety manager to set up such a change communication process. The product line system safety process is not managed and maintained by a safety manager in the product line; the safety process is monitored by the central assessments and audits group. Normal periodic employee/manager communication is supplemented with reviews of change requests with the engineering manager. This might be periodic or ad hoc: for example, whenever a change comes to the attention of the engineering functional manager that is recognized as perhaps having a safety impact to be examined by a safety expert. Change communication is less systematic. Audits by the central system safety organization may drive systematic change reviews by the safety expert if a finding needs to be escalated. Such an escalation can be raised by the safety auditor as an issue requiring action by the product line leadership to resolve.

In alternative 3, there is a dotted line relationship between the system safety manager and the product line. While the safety manager does not directly report to product line leadership, the performance of the product line system safety manager affects metrics reported to senior executive management concerning execution of the safety process in the product line. This relationship motivates closer cooperation and support between product line leadership and the product line safety manager. The system safety manager for the product line does not directly control the product line system safety process but can ask to be included in the review of potentially safety-related change requests, and can influence making this part of the product line process. Such a request can be supported by the central system safety auditing and assessment organization through monthly reports or audit reports. Having such a process is audited internally and perhaps externally.

It is a fundamental need to sustain the safety of the products launched. Such changes are a vulnerability of the safety process if they are not systematically reviewed for safety impact. ISO 26262 requires this for functional safety. Any external auditing organization or customer will review this, and compliance is expected. All three alternatives can comply.

Communication channels among different product lines within the enterprise is also affected by the organizational alternative. These communication channels are with respect to system safety issue resolution and best practices. Significant differences in these practices can adversely affect safety cases. In alternatives 1 and 2, this communication can occur by means of the central organization sharing observations from other assessments and audits. Issue-resolution practices can be observed as findings are resolved from audits, or as issues are resolved and observed with respect to the product during the duration of the program. This communication can be conveniently implemented during periodic audits, since the auditor typically audits multiple product lines. If there are significant differences that adversely affect a product's safety case, then there may be a finding to correct the practice that is then reviewed during a subsequent audit. Failure to implement this practice by this subsequent audit triggers an escalation by the auditor to product line leadership for support in implementing the improved practice.

As similar issues arise in different product lines and need to be resolved, the auditor can share with other product lines approaches that have been successfully implemented.

Contacts with other product lines may be recommended or established to facilitate resolving the issues. Such contacts can further improve communication concerning system safety among the different product lines. In addition, the central system safety organization can host global workshops to facilitate sharing of best practices among all the product lines within the enterprise. Besides the sharing of best practices, such workshops provide an opportunity for informal communication and networking among the safety experts and other engineering personnel and management who attend the workshop.

In alternative 3, the product line safety manager is a member of the central system safety organization. Normal day-to-day activities within the central system safety organization require communication among the organization's safety experts. Weekly or monthly meetings provide opportunities to learn what is being considered and what actions are being taken concerning system safety in all of the enterprise's product lines. In the organization, the product line safety managers are colleagues and can share best practices and experiences directly. In addition to learning about what happened or may happen, discussions can promote synergy among the safety experts in determining what should happen as a best practice within the system safety organization before the practice is implemented. Safety experts can be rotated among the product lines to systematically implement best practices and to broaden their domain knowledge. This rotation increases not only their knowledge of different product lines but also their network among the engineers and managers of the product lines who perform domain-specific safety analyses and produced other safety artifacts. They can consult at and assist with the other product lines for which the product line safety managers provide support. This facilitates communication among product lines not only because of the safety expert's technical knowledge, but also because of their personal contacts in other product lines. Further, rotation helps prepare product line safety managers for roles of assessment, auditing, and other system safety management positions. This networking can multiply the communication channels, and thus information flows more naturally.

Every organizational choice is a trade-off of advantages and disadvantages. These organizational choices are in both the hierarchy of the enterprise as well as the management of system safety personnel. Despite what appear to be clear advantages for the enterprise, the organization of alternative 3 may be unpopular with product lines: they may not want to relinquish solid line reporting of the safety experts. System safety leadership may be reluctant to propose this organization because it appears to be self-serving, or a "power grab." The advantages of alternative 3 for the execution of the system safety process may be too politically toxic to implement in the enterprise.

In this case, alternatives are required, to optimize an achievable organization that can be implemented for system safety. If the enterprise is not structured by product line but instead by some other attribute or function, this may change the associations of the safety experts to entities of the enterprise but may be analogous to the three alternatives presented in this chapter. Other alternatives are possible: for example, consider a mixture of two or more of the alternatives presented in this chapter, embraced by a single enterprise. While one alternative may have advantages, another alternative may be more convenient to implement because of other priorities in the enterprise. Product line leadership may assign a greater priority to some engineering domains than others, or some engineering domains may be assigned to be managed by or combined with other engineering domains.

For example, consider alternative 1: product line leadership has direct solid line management responsibility for system safety for the product line. Product line leadership may be uncomfortable with a direct report from system safety. Product line leadership personnel may not have had an opportunity to gain experience managing system safety functions. The remaining domains, other than system safety, may be complex and time consuming to manage, with demanding issues to resolve such as meeting customer demands and ensuring deliveries. There may already be a report from the engineering quality organization supporting automotive software process improvement and capability determination (SPICE). Compliance with the demands of the base practices of SPICE is demanding and time consuming, but not directly related to a customer deliverable in most cases.

Where does it end? Can any of this extra reporting be delegated to another direct report to unload the product line leadership? The product line system safety manager may be assigned to report to another discipline, such as system engineering, that already has a seat at the product line engineering leadership table. Thus, the product line engineering executive avoids another direct report. Mentoring of the product line safety manager may still be provided by the central system safety organization. The product line safety manager may explain the need to establish or improve the product line system safety process to the systems engineering manager, and win support for proposing to product line leadership that the improvement be implemented across the product line. This is necessary because in this example, system safety management does not have a seat at the product line leadership table. Nevertheless, support may still be provided through the central system safety auditing and assessment organization by means of metrics supported by comments, audit reports, and other periodic system safety reports.

The previous discussions may help identify weaknesses in different variations of the three alternatives presented. The five critical success factors are important criteria for these evaluations of alternative organizations: failure of an alternative to consider and enable these factors may lead to unresolved issues. In these cases, measures may be taken to mitigate weaknesses due to trade-offs made. This may require altering the organizational structure or implementing a unique career path for system safety personnel. Ownership may be assigned for this mitigation, and continuous improvement is possible.

3

System Safety vs. Functional Safety in Automotive Applications

Safety Terminology

The word *safety* is used in many different contexts in the automotive industry. For example, *safety* may be used when referring to vehicle occupants as their freedom from harm in a vehicle collision; this may be simulated by the behavior of and stress imposed on crash dummies of differing types. Requirements for product safety are included in IATF 16949, a quality standard, but product safety is not defined in this standard. The reader is left to interpret the intent of the standard for product safety based on the requirements that are included.

Safety systems are included in vehicles and are often described as *passive* or *active*. This distinction is made between systems that function after a collision has occurred and systems that "actively" function to prevent a collision or reduce the severity of the collision. Passive safety systems include seatbelts and airbag systems. Seatbelt systems are standard and can include advanced features such as pretension functions and seatbelt buckle lifters. Active safety includes advanced emergency braking and lane departure warnings. There are more automated features that fall into the category of advanced driver assistance but that are also considered active safety systems.

Functional safety refers to the safety of the system when it malfunctions. Some enterprises have functional safety organizations with responsibility for the functional safety process. While the term *functional safety* has a limited scope, including only safety when there is a malfunction, sometimes the term is misused with a broader intent to include safety without a malfunction. *Product safety* is sometimes used to distinguish this safety without a malfunction from functional safety as described in ISO 26262. This term is used when referring to the safety of a system without a failure and, broadly, includes the safety of the intended function (SOTIF) as well as the physical safety of the system, such as the absence of burrs and sharp edges and the safety of the human-machine interface when operating as designed. *System safety* is used for the combination of functional safety and product safety, including SOTIF, and is a broader term.

Automotive System Safety: Critical Considerations for Engineering and Effective Management, First Edition. Joseph D. Miller.

Functional Safety Standards vs. System Safety

Background

Since 2011, functional safety in the automotive industry has been highly connected to ISO 26262, which was revised in 2018. This functional safety standard has had broad publicity in the international safety community and the automotive industry in general. There have been many conferences every year since its release as well as numerous courses based on this standard. Even though the standard itself does not endorse certification, there are certification courses for ISO 26262.

Prior to the first release of ISO 26262, some companies applied IEC 61508 or a military standard to automotive product development. A standard such as IEC 61508 considers the various phases of a product's lifecycle. These phases can be aligned with the phases of automotive development, and the standard's requirements can be adopted by company's automotive product development process. Companies did this using their own interpretation of the standard or the interpretation of a safety consultant. In so doing, the benefit of being able to show compliance with a safety standard was achieved before the automotive functional safety standard was agreed on and published. The goal was to ensure an acceptable level of risk if the product malfunctioned, and to generate evidence of diligence that this was accomplished. Evidence of fulfilling the requirements of a standard supported this goal of diligence. The documentation of fulfilling these requirements, even if there were no mandatory work products, helped show the care that was taken concerning safety.

The IEC 61508 and ISO 26262 standards differ in their approach to functional safety. IEC 61508 was intended to be a standard that could be followed in almost any industry or be the basis for a standard in that industry. ISO 26262 is intended to be a derivative of IEC 61508 that is dedicated to road vehicles. Table 3.1 illustrates the differences between the two standards.

Application of Functional Safety Standards

Using either of these standards can provide guidance for achieving functional safety for an automotive product. Both have lifecycles that can be tailored to the needs of a particular automotive product being made ready for launch. To use IEC 61508, the lifecycle needs to be tailored to a serial production product. This is different from a non-automotive product that is built and then has safety mechanisms added to it, like a manufacturing facility or a power plant. The safety mechanisms need to be included in the design and verified before series production and public sale. Then the analyses of what hazards are applicable, and how they may be caused by a malfunction, can be used to determine functional safety requirements. Such systematic analyses to elicit requirements can provide confidence that the safety requirements are complete. Likewise, using ISO 26262, similar activities are pursued to elicit functional safety requirements. There are some prescribed safety analyses, like determining the hardware architectural metrics to elicit requirements for the safety mechanisms to detect hardware failures; evidence of compliance completes the safety case. This evidence has to be traceable to each individual safety requirement to provide evidence that no requirement was missed. All other requirements of the standard require evidence of compliance, and the work products help with this.

Table 3.1 Comparison of IEC 61508 and ISO 26262.

IEC 61508	ISO 26262
• Applies to electrical/electronic/programmable electronic systems (E/E/PESs)	• Applies to safety-related systems including E/E systems
• Is generic and applicable regardless of application	• For series production vehicles
• Applies to systems without application-sector international standards	• For series production vehicles
• Has quantitative goals for avoiding hazards based on safety integrity level (SIL)	• Does not have quantitative goals for avoiding hazards
• Lifecycle implies the system will be built and then validated, like a chemical or nuclear facility	• Uses the automotive lifecycle such that the item is validated before series production
• Uses SILs 1–4, based on risk to index the requirements	• Has automotive safety integrity levels (ASILs) A–D, based on Severity (S), Exposure (E), and Controllability (C), to index the requirements
• Does not have explicit work products	• Has over 100 normative work products

However, compliance with ISO 26262 does not necessarily lead to achievement of system safety for automotive products. System safety includes more than evidence that (i) the requirements of ISO 26262 have been met, (ii) the requirements have been elicited for safety of the product during a failure, and (iii) these requirements have been complied with. Elicitation of requirements for safety when the automotive product is in use without a failure must also be considered. Safety of the product in use without a failure is specifically out of the scope of ISO 26262: this is not part of functional safety, because functional safety only includes safety in case of failure. It is not the intent of the functional safety standard to provide guidance for eliciting safety requirements when there is no failure. System safety *does* require consideration of the safety of the product when there is no failure. For example, for an automatic emergency braking system, a failure can cause unwanted emergency braking that could be a hazard for a following vehicle. This same unwanted emergency braking could be caused by limitations of a sensor or an algorithm without a failure. It is necessary to achieve functional safety to achieve system safety. ISO 26262 compliance is not enough – system safety requires more.

Still, organizations that do not separate functional safety from system safety can use the functional safety standard for guidance. The guidance provided for functional safety can be generalized beyond the scope of the standard and used to consider other causes for "malfunctioning behavior" besides failures of hardware and software. The same malfunctioning behavior could be due to system limitations without a failure. Analyses can be performed to determine these limitations and ensure safe operation of the system. Hazards include the personal interface with the system when operating properly: confusion or misinterpretation of system capabilities can be eliminated through thoughtful design, including mitigation of foreseeable abuse.

Cuts and bruises as well as hazards during maintenance when the system is not in operation, such as those caused by the release of stored energy, should be considered. Such considerations result in safeguards and warnings as well as maintenance procedures, such as deploying the airbags when disposing of the vehicle. Evidence of analyses to elicit these requirements is included in the work products, which also contain the evidence required by the work products of ISO 26262. This can be an extension of the functional safety evidence to include system safety evidence. ISO 26262 is useful for achieving system safety, but it is not sufficient.

Safety of the Intended Function (e.g. SOTIF, ISO PAS 21448)

Safety when the automotive product is in use without a failure is an especially important consideration for an advanced driver assistance system (ADAS) and automated driving. The Organisation Internationale des Constructeurs (OICA) and Society of Automotive Engineers (SAE) standard J3016 define ADAS and different levels of automation. There are six levels, 0–5:

1) *Level 0* is defined as no automation. This may include a steering system or a braking system.
2) *Level 1* is driver assistance and has either assisted steering or assisted acceleration/deceleration. This may include lane-keeping assist or automatic cruise control, but not both.
3) *Level 2* is partial automation and has both assisted steering and assisted deceleration. This may include lane-keeping assist and automatic cruise control, perhaps with automatic emergency braking.
4) *Level 3* is conditional automation and automates the dynamic driving task with the expectation that the driver will take over if asked. This can include an automated driving feature used on the highway, but the driver may take over under some adverse conditions or in case of failure within a reasonable time.
5) *Level 4* is high automation and automates the dynamic driving task for specific driving mode performance without the expectation that the driver will take over. In conditions like those outlined for Level 3, the system must reach a minimum risk condition on its own or have sufficient redundancy and reliability to complete the driving task
6) Level 5 is full automation and automates the dynamic driving task full time without the expectation that the driver will take over.

To achieve system safety, these types of automotive products need to be safe in use as well as safe in failure. They are in scope of ISO 26262 with regard to functional safety. All the requirements of ISO 26262 are applicable to these systems; this was a major consideration when the second edition of the standard was being developed. Significant guidance is included in ISO 26262 about fail-operational considerations.

Up to level 2, ISO PAS 21448 is intended to provide guidance for safety in use. The scope is limited because the experts working on this publicly available specification (PAS) felt that the maturity of more-advanced products in the field would not support more-definitive guidance at the higher automation levels. Nevertheless, the first edition of ISO

PAS 21448 states that its guidance may also be useful for higher levels of automation. ISO PAS 21448 is expected to be referenced, as the requirements have been vetted by a team of safety experts knowledgeable in the domain. When the standard based on ISO PAS 21448 (ISO 21448) is released, it is planned to provide guidance for ensuring the SOTIF of higher levels of automated automotive products: it is expected to include normative goals for the achievement of the SOTIF.

The current ISO PAS 21448 includes some new definitions to support a SOTIF-related hazardous event model, including an associated taxonomy for scenes, events, actions, and situations in a scenario. The definitions from ISO 26262 are also included in ISO PAS 21448 by reference (a statement is included in ISO PAS 21448 that the definitions of ISO 26262:2018-1 apply to ISO PAS 21448). This helps maintain the consistency of language for both functional safety and the SOTIF in road vehicle safety. This taxonomy supports both initial analyses as well as verification and validation of the automotive product for production. Analysis based on this SOTIF related hazardous event model strives to consider all the relevant use cases in which the function may be used. Much of this consideration concerns triggering events, which are key to determining the SOTIF. This will be discussed next.

Triggering Event Analyses

Background

Triggering events are important because they serve as initiators for any reaction by an automotive system. Automotive systems that react automatically do so as a result of external conditions that are perceived through sensors. The automotive system processes these perceived conditions with algorithms that determine whether or not to react, and if so, how to react to the conditions perceived. The reaction is due to conditions in a driving scenario, so these conditions in the scenario are the triggering event.

For example, consider a lane-keeping assist system that commands an electric steering system with steering torque or a steering angle to position the vehicle laterally in the lane, and a camera to sense the lane boundaries. This lane-keeping system begins to provide this command as the camera perceives that the vehicle is starting to leave the ideal position in the lane and approach the lane boundaries. A triggering event might consist of conditions where

1) The lane boundaries are clear enough to be sensed unambiguously. For example, there is only one lane boundary marking on each side of the lane, there are no other redundant markings, and the color of the lane boundary falls within the bandwidth of the camera.
2) The turn signal is not engaged so no intent by the driver to leave the lane has been indicated.
3) The vehicle is positioned laterally such that it is about to cross the lane boundary, and there is no contradictory information or other ambiguity about the vehicle's position.

This event triggers the camera to send a signal to the steering system to start correcting the vehicle's lateral position. It is perceived that the vehicle may leave the lane when it is not the intention of the driver to do so. There is no other indication that this perception is incorrect or unreliable.

This is one triggering event. The conditions perceived trigger a reaction from the system that the system has determined to be justified because the system is operating correctly and there is confidence in the perception. If this is done correctly, and there are no failures, the vehicle should start to move to the center of the lane before it can leave the lane. This is the intended function of the system when it is operating as specified, in order to avoid unintended lane departure. As this is a correct reaction of the system design, it is a known safe scenario per ISO PAS 21448: area 1 contains known-safe scenarios, area 2 contains known-unsafe scenarios, and area 3 contains unknown-unsafe scenarios. The scenario discussed previously goes to area 1.

Now consider the same system, with revised conditions:

1) The normal lane boundaries are clear enough to be sensed, and additional construction lane markings are clear enough to be sensed. For example, road construction lane markings are a different color, but that color is within the bandwidth of the camera being used.
2) The turn signal is not engaged, so no intent by the driver to leave the lane has been indicated.
3) The vehicle is positioned laterally such that it is about to cross the lane boundary of the normal lane but stay within the construction lane markings, guided by driver input torque, and there is no contradictory information or other ambiguity about the vehicle's position or the driver input torque.

This event could trigger the camera to send a signal to the steering system to start correcting the vehicle's lateral position because the normal lane markings are clear, and the vehicle is about to cross them with no indication from the turn signal that this is the driver's intent. In that case, the vehicle would start to move to the center of the normal lane before it could leave the lane. Such an action by the system falls within the specified behavior of the system, and it may be a correct reaction, to correct a driver error. However, the correct reaction might also be to do nothing and continue to follow the construction lane markings: this could be considered the driver's intent based on the driver's controlled torque input to follow the construction lane markings. A third possibility would be to switch off the lane-keeping system and notify the driver that the system is being disabled because the system is unable to determine the appropriate response in this scenario. In that case, the driver would continue to operate the vehicle without the assistance of the lane-keeping system.

These conditions are another triggering event. The correct response needs to be determined and specified so that it is repeatable and so can be evaluated in other, similar scenarios. This scenario may be classified as a safe scenario after the appropriate reaction is determined and the evaluation is favorable. However, for now, for this system, let's assume the reaction is indeterminate and a safe response may not be repeatably achieved now even though such a response may be repeatably achieved later. This becomes a potential unsafe scenario per ISO PAS 21448, and it goes to area 2.

Systematic Analyses

This part of a triggering event analysis needs to be done systematically for every scenario the automotive system may encounter that may have a triggering event that could lead to an unsafe reaction, including unique scenarios that are only encountered in some countries where the system is sold. For example, if a lane-keeping system is to be used in Korea, the analysis must take into account that blue lines are used to demark lanes used by buses. Other markings inside lanes need to be dealt with by various systems, such as advanced emergency braking, because these markings vary from country to country and have nuanced differences in appearance when approached from different elevations.

While completeness is difficult to prove (for example, what if a black swan sits on the camera?), by proceeding systematically, confidence in the completeness of the analysis increases. Each of the relevant road types, environmental conditions, possible relevant vehicle interactions, maneuvers, road users, routing exchanges, intersections, and so on need to be evaluated and searched sufficiently for unknown scenarios with evidence of diligence. Guidance for determining the sufficiency of such evaluations can come from ISO PAS 21448 or from another source, such as a specialized consultant. If portions of the specification are carried over from a previous product that has been verified and validated, they can be used to tailor the current product's verification and validation using an acceptable rationale.

For example, consider an automated emergency braking system that has already been validated using fleet data and simulations on a particular vehicle. The specification is unchanged except for the system's operation at very low speeds. In this case, after a nominal verification of the existing data to be carried over, the remainder of the validation might be focused on the potential triggering events for the changed lower-speed scenarios. To achieve a systematic triggering event analyses, first consider the hazards that may be caused by the system, such as errant steering assist by a system that commands steering assistance. Using a hazard and operability study (HAZOP) technique with guidewords (e.g. *too much, wrong timing, in error, missing*) concerning the actuator authority and external communications of the system is a systematic way to do this, similar to ISO 26262 or the explanations in SAE J2980 [6].

Then the potential driving conditions where the system will be used are listed (e.g. highway including entrances, city), and all the potential conditions are considered systematically. There is considerable scope to these considerations, which can be considered in multiple categories as hinted previously: for example, speed range, traffic conditions, environmental conditions, infrastructure including variations due to location (such as the blue bus lane markings in South Korea), lighting conditions, and special conditions such as tunnels and bridges. Knowledge of the system's sensor limitations or other prior knowledge helps prioritize the scenarios and conditions to be evaluated.

Behavior of all the actors must be considered, such as the driver (fatigued, startled, distracted, etc.) and other road users and vehicles (e.g. bicyclists turning, walking their bike, and perhaps falling; trucks crossing, including semi-trailers and full trailers of different lengths; emergency vehicles passing; and passenger cars stopping). Each of these needs to be considered in multiple slightly different scenarios. Behavior with respect to sensor limitations (radar range, camera signal to noise ratio, etc.) must be considered, using simulations in combination with road data for efficiency if validity is confirmed.

Potential misuse must be considered (e.g. failure to take over when signaled) and also simulated, if validity is confirmed, to gain additional efficiency and scope of the scenarios evaluated. Misuse scenarios do not consider abuse, although abuse can be addressed by system features if available.

All these considerations must be taken together and in combination. For example, misuse scenarios should be considered in combination with intersections where large trucks are crossing, if this is relevant for the system under evaluation. Other combinations may be considered if they are at least plausible to be encountered by the system, even if exposure to such combinations seems remote. This grows into massive analyses of just items that can be foreseen, as required to populate areas 1 and 2 of ISO PAS 2144 in figure 8 in this standard. Preparing to evaluate all known potential triggering events requires extensive detailed compilation and organization. ISO PAS 21448 provides further guidance in clauses 6 and 7 as well as the annexes; this guidance is useful for understanding the activities needed and information that is useful for evaluating triggering events. Guidewords and examples are also provided, along with foreseeable misuse guidance, which help to construct a systematic approach that can be managed.

Constructing a spreadsheet or database containing potential hazards helps to formulate a systematic approach to the evaluation. First the different driving conditions are considered for each potential hazard, to determine if they are applicable. Some driving situations are not applicable for every system to be evaluated, due to internal system restrictions or geofencing of the application. For example, an automated driving system used to transport passengers on the grounds of a business or other entity may have no high-speed conditions to be evaluated. So, high-speed conditions can be removed from consideration when organizing the evaluation for this system.

The matrix or database grows as each category and each condition are considered. Such conditions include speed, traffic, wet roads, fog, etc. are considered individually as well as with plausible combinations (e.g. speed and traffic). If a combination is implausible, it can be removed. Possible environmental conditions are overlaid upon all of these: for example, driving in traffic during fog or rain, or on icy streets. Sensor limitations are added to each of these, along with other road users and their behavior: following a motorcycle that changes lanes often, pedestrians in a college town who jaywalk frequently, and other commonly foreseeable behaviors. Potential misuse must also be considered for each case: for example, not holding the wheel and driver distraction when engaging adaptive cruise control and lane keeping while the motorcycle is changing lanes; this causes braking authority to be required outside the capability of the cruise control system.

These analyses can be executed more efficiently using a simulation, which provides data for the combinations of scenarios and conditions so the most relevant are considered further. This approach also improves accuracy and completeness. The simulation can more accurately represent the situation that is evaluated, and variations of the conditions used in a scenario that is otherwise unchanged can be directly implemented and reviewed. This review is required not only by the safety organization, but also by experts from engineering. They may observe flaws in the scenario or flaws in the intended system behavior that can be corrected and repeated for evaluation. As with other analyses, findings lead to actions. Corrections needed in sensors or algorithms can be made or simulated and reevaluated. Limitations of system behavior can be avoided by limiting the scope of the intended functionality. ISO PAS 21448 discusses such specification changes: this step is included in the flow after evaluation of each of the areas, if required, and improves the product.

In addition, further triggering events can be sought by actual road use, which may potentially result in updated simulation. Modifications can be made in the simulation to construct scenarios that are not intuitive or that result from the systematic analysis, and thus find triggering events that fall into region 3. Often this is done with fleets that are running during development. It is common in automotive development to have fleets of production intent vehicles driving in different environments for extensive evaluation on public roads before series production and public sale. The systems are turned on, but the actuators are disabled to avoid potential hazardous behavior by the vehicle. The existence of such potential hazardous behavior may be discovered by reviewing the data or through an alert or record kept during the fleet testing. Multiple systems can be evaluated at the same time on the fleet vehicles: for example, advanced cruise control, automated emergency braking, and lane keeping may all be installed on each vehicle for evaluation during fleet testing. Scenarios may be sought in the planned routes for mileage accumulation by the fleet to evaluate scenarios of significance for each of these systems; this also allows their interactions to be monitored. Arbitration may be required for systems that engage the same actuators, so the correct priority is executed. For example, consider a lane-keeping system and a system that uses steering correction to provide greater traction when an advanced braking system is decelerating on a split mu (coefficient of friction) surface. The steering commands may be in conflict, so priority must be determined. During fleet evaluations, the behaviors and sensor outputs can be recorded, possibly including video recordings to enhance understanding during evaluation and validation after the testing. We will discuss validation next.

Validation

Validation of the SOTIF for public use follows from the triggering event analyses. The system has been evaluated through extensive review of scenarios, and any needed improvements have been completed. The goal is to validate that the automotive system will not cause an unacceptable risk to the general public, as well as to show compliance with governmental and industry regulations. This requires a criterion to be used to evaluate whether the risk to the general public is acceptable or not. To do this, validation targets and a verification and validation plan must be developed. There are practical considerations for this validation plan, so constructing the plan must take into account a method to reach the validation targets.

Validation Targets

Validation targets vary based on the validation methods chosen, such as mileage accumulation, simulation, and analyses. If a sufficient rationale is provided that validation can be completed by analyses, then a criterion may be that no potential triggering event leads to a hazardous event. The analyses can confirm that no violation of this criterion is possible.

If fleet driving is the validation method chosen, then a number of miles with no hazardous event may be chosen based on available traffic statistics for the possible types of accidents. The miles between potential hazards created by triggering events must exceed the number computed from traffic statistics for the average miles between the same plausible accident type. Sometimes a factor greater than one is used to increase the target to a mileage much higher than that computed from traffic statistics, in order to be conservative or to mitigate the *dread factor* and support market acceptance. This dread factor is similar to what occurs in other forms of transportation, such as flying: when there is an accident, people recoil from flying even though flying is one of the safest means of travel. To avoid this recoil, the risk of the means of transportation or automated feature needs to be negligible compared to what has already been accepted through everyday use. This is not a valid criterion for a reasonable level of risk – risk is actually reduced through the use of the automated feature. Nevertheless, such a goal is aspirational for business reasons.

Efficiency and confidence can be gained by a combination of methods, each with validation targets. Equivalent mileage targets can be computed for simulations of scenarios that are difficult to encounter in driving or that are synthesized, such as combinations of conditions or journey segments through appropriate simulation algorithms for automated selection. Then confidence gained from analyses and demonstration is taken into account so that the level of confidence required from actual miles driven is reduced, thus reducing the required mileage to be accumulated to an achievable level. Expert judgment is also taken into account to reduce the required mileage to be accumulated and may be necessary for a practical validation plan. Such mileage reduction requires an acceptable rationale.

ISO PAS 21448 envisions this pragmatic approach. Multiple examples are included that allow some alternatives to be considered and that support similar rationales to ensure that acceptable validation is accomplished. The release of systems to implement the different levels of automation is expected to improve automotive safety significantly. This adds momentum to the objectives of validating and releasing these systems to benefit the safety of the general public. The safety improvements these systems provide can mitigate some of the risk in releasing these systems: the potential hazards can be outweighed by the potential benefits. A similar argument exists for seatbelts and airbags. Seatbelts potentially keep an automobile occupant from being thrown clear of an accident, while airbags can cause injury to an occupant because of the force of their deployment. There is some risk in using these passive safety systems, but the reduction of risk in an accident far exceeds this danger. Nevertheless, due to the risk of bad optics automotive suppliers and vehicle manufacturers (VMs) seldom make these trade-offs: VMs and suppliers tend to view that such arguments, even if valid, may be seen to be self-serving, and therefore take a more conservative approach or leave the rationale for regulatory agencies to determine. It is not uncommon for regulatory agencies to study this risk-versus-benefit trade-off as they did for airbag systems. They do not stand to gain financially from the deployment of systems that improve safety of the general public, and their goal is overall safety improvement.

Requirements Verification

Prior to validation, verification of system and subsystem requirements is helpful. This is an expected system engineering practice, but it is especially important when considering system safety. When validating that a design meets its validation targets for safety, it is important to know that the design is correctly implemented. This can be achieved by systematically eliciting requirements and verifying compliance with them. The sources of requirements will be discussed in a Chapter 8 of this book.

Sensors deserve special consideration. Requirements for sensors, and verifying compliance with these requirements, are key to understanding the system's capabilities and limitations, and to finding and evaluating potential triggering events to ensure the SOTIF. In order to validate that the system is safe under conditions at the sensors' specified performance limits, first verify that the sensors achieve these limits. This is important for evaluating the system through simulation or testing. Then, for example, artificially increase the signal to noise to elicit a response for evaluation. Other simulation methods can be used to evaluate behavior at the limits of the sensor's capability.

A verification plan needs to consider all the requirements specified. Traceability is needed to verify that every requirement has been verified, including algorithms and actuators. Such planning needs to include verifiable pass and fail criteria for algorithms. ISO PAS 21448 discusses this verification along with integration testing.

The three areas discussed earlier are also important for validation. Additional analysis of triggering events continues into validation to identify or resolve findings. Track demonstrations can verify that findings have been resolved as intended, and additional road testing can show that unknown triggering events do not cause unsafe system behavior. Failure of validation during (i) analysis; (ii) demonstration in off-road conditions, like a proving ground; or (iii) on-road testing requires improvements to be made in order for validation to be successful. Repeating the analysis and demonstration can confirm that improvements are effective. Therefore, the flow is expected to be first to area 1, to evaluate the triggering events; then to area 2, to create a verification and validation plan and evaluate the known triggering events; and finally to area 3 to evaluate unknown triggering events. This flow is required so that work proceeds efficiently. In addition, off-road testing is needed to demonstrate diligence in considering the safety of the general public with documented, successful, proving ground results. Targets are set to establish success criteria for these validations. Risk of harm to the general public must not be unreasonable.

The safety requirements are to avoid hazards that may be caused by the intended function, so these hazards must be determined. First, a specification is developed and then a hazard and risk analysis is performed. A specification, however brief, is required first (like an item definition for ISO 26262), so the boundaries of the system are established and its functionality is clear. If the risk is acceptable, further system safety analyses and safety validation are not needed. For example, if the system's authority is not sufficient to cause a hazard, perhaps because another independent system monitors and limits it – such as a steering system limiting the commands it receives from a camera to a safe and controllable level – then the risk from the system is not unreasonable. If potential hazards with an unacceptable risk exist, then the triggering event analyses, as described earlier, are performed in area 1. This triggering event analysis is sufficient to cover the scope of functionality intended by the specification in all relevant use cases that may arise after the system is deployed. Each case is reviewed and evaluated to elicit any unacceptable risk. Not all unintended behavior is an unacceptable risk: unknown safe behaviors may be discovered. If any are found, then the system specification is revised to incorporate performance improvements or limitations that preclude the unacceptable risk, such as a reduction in capability. If no further specification changes are required, then verification and validation start in area 2.

The process of verifying sensors and subsystems is planned and performed as mentioned previously. Sensors and subsystems must be specified in sufficient detail so that their intended limitations are understood. Verification confirms that their capability, at a minimum, achieves these limits.

Then the system's response to triggering events is validated. Knowing that sensors and subsystems achieve the intended performance limits enables validation, taking into account potential triggering events that fall within these limits as well as potential triggering events that fall outside the limits, such as distant objects, stationary objects, or faint or obscured lane markings. Some validation can be performed in simulation, which allows systematic degradation of sensed input targets to determine whether system performance remains safe as specified limitations are crossed. Actual vehicle testing is often employed as well, and may overlap and repeat some of the tests performed in simulation. These repeated tests also serve to validate the model. Evidence that the model has been validated strengthens the safety case and adds confidence to the simulation results.

Triggering events and scenarios can be grouped into equivalence classes to reduce the number and scope of scenarios to be evaluated and thus improve efficiency. Often, boundary cases are referred to as *edge cases* or *corner cases*. Edge cases are at the limit of a single sensed parameter, and corner cases are at the intersection of the limitations of multiple parameters. By validating edge cases and corner cases, an acceptable rationale can be developed for accepting that the entire equivalence class is validated. Such evaluations also strengthen the safety case, because they provide evidence that the "worst cases" have been considered.

Identifying edge cases is also useful when machine learning is employed to train artificial intelligence (AI) used in automotive safety systems. Efficiency and consistency are gained by reusing the analyses used for validation. Edge cases and corner cases can be used to train a machine and thus incorporate validated behavior into the machine. Other members of the equivalence class can be used for verification, which also confirms the assumptions used to justify the use of equivalence classes. This verification of the AI algorithm is performed prior to validation.

In area 3, unknown triggering events are sought. This is accomplished primarily by fleet testing with a strategy and bias to explore use cases that were not uncovered in analyses, track testing, or simulation. If a target mileage without a hazardous event is the goal, it can be determined directly from traffic statistics or modified using data from previous systems or expert judgment based on the severity and rigor of the test plan. This target mileage is achieved by exercising the safety system in real-world situations and proportionately evaluating it in all the driving conditions, locations, and environmental conditions and situations in which it is specified to operate, as well as those outside specification that are reasonably foreseeable. A fleet can be allocated proportionally to these regions and the mileage agglomerated to reach the overall target mileage more efficiently. If there is an anomaly, corrections are made to remove the root cause, and then the mileage is counted.

For example, suppose one-third of the necessary mileage had been accumulated toward the target, with no triggering event causing an unsafe response by an automated emergency braking system. Then an unwanted emergency braking event occurs. Upon analysis of the root cause, it is discovered that the field of view allows a reflection that, under a rare set of conditions, triggers this response. The specification and actual field of view are adjusted to remove the possibility that this could ever happen again, and none of the intended function is altered. Now, rather than starting over and ignoring the accumulated data, an acceptable rationale is to count the previously accumulated miles and continue fleet testing with the modified system. This is an acceptable action because it continues to protect the general public from unacceptable risk, and it complies with ISO PAS 21448. If there is a concern that the change may adversely impact previous evaluations, recordings of these previous situations can be input into the safety system to reconfirm the previous validation mileage. Documentation of this action will further support the rationale for accepting the corrective action.

Simulation also helps accumulate mileage but is not sufficient by itself. A combination of simulation and actual driving may be justified by a rationale that defines the purpose and goals of both. Simulation allows the testing of plausible situations that are difficult to encounter in real life, or that can be simulated with greater safety than by driving. However, unknown situations can result from scenarios outside the scope of the simulation, and whereas simulation may not generate such situations, driving will.

Release for Production

ISO PAS 21448 provides guidance and a flow chart for determining whether the SOTIF has been sufficiently validated to recommend release of the automotive product. The committee directed this author to produce this flowchart, included as Figure 14 in ISO PAS 21448, which summarizes the text of clause 12 of the PAS, discussed next. First, there needs to be evidence that the validation strategy took into account all of the use cases. Failure to consider a relevant use case could reduce confidence that the requirements are complete, and that compliance is demonstrated. The strategy discussed previously ensures this is done as practically as possible. Combinations of methods are employed to ensure confidence that no potential use case is neglected. There is confidence that this strategy is complete because of the methodology of eliciting triggering events. A compelling rationale can be documented to justify this confidence. Then the triggering events are systematically validated either directly, as a member of an equivalence class, or from verification in a previous product. Expert judgment may be considered because of the rigor of emphasizing perceived worst-case scenarios through the selection of use cases to be evaluated. The evidence can be analyses or reports that include the rationale.

Second, there needs to be evidence that the product achieves a minimum risk fallback condition that does not impose unreasonable risk on the vehicle occupants or other road users. This minimum risk fallback condition varies, depending on the level of automation, from shutting down the system with a driver notification to managing the fallback internally with no expectation of driver intervention. Again, the systematic approach previously described should provide this evidence by systematically examining scenarios, including scenarios beyond the sensors' specification or foreseeable misuse by the driver. The automotive safety system may then be triggered to enter a fallback condition, which must achieve a minimum risk condition that is acceptable based on regulations as well as the reasonableness of any risk imposed on the general public. Only specified driver interactions are permitted: these may include taking over in some period of time, or no action. It is expected that the driver is warned when such a fallback condition is engaged. This may allow takeover even if none is required by the system. Then the driver takes responsibility for operating the vehicle safely instead of the minimum risk condition of the system. This system fallback strategy ensures minimum risk, and the transition is recorded.

Third, sufficient verification and validation need to be performed to provide evidence that no unwanted behavior will be triggered in any scenario that will be encountered after launch. This can be accomplished by systematically eliciting and verifying requirements and achieving the validation targets established in the validation plan. Again, evidence of completeness, as described previously, is key to this rationale. Proving completeness may not be possible, but the systematic methods employed give confidence of this completeness. Since the system is verified first to ensure that the specification assures no unwanted behavior due to any triggering event, and there is confidence in the completeness of the requirements in the specification, there can be confidence in the validation. This is because the validation is based on the specification, and the validation targets are set so that the specified functions are executed safely in all relevant use cases. Then the validation targets are met sufficiently to ensure no unreasonable risk, thus completing the validation successfully.

When all of this has been completed, the product is recommended for release with respect to SOTIF. Any unexpected anomalies have been resolved by removing the root cause from systems intended for series production and public sale. No unreasonable risk has been discovered, and the evidence is sufficient.

If the triggering event analyses, verification, and validation discover no unacceptable behavior, but completion of the validation testing for unknown triggering events requires more time to meet the target, a conditional release may be possible. Such a conditional release requires a judgment that that there is enough evidence from the uses cases of greatest priority to provide confidence that the validations will be met successfully. This varies from one system to the next and is more restrictive as the level of automation and the scope of usage increase.

If there is an unresolved unacceptable behavior, then release may not be recommended. The recommendation for release in this case may first require a root-cause analysis of the issue, resolution, and permanent corrective action. The corrective action may improve the performance of the product or prevent exposure to the scenario in which the unwanted behavior occurs, such as a vehicle speed limitation that disengages the product. Verification that this corrective action permanently resolves the issue is needed. A rationale must be provided for a practical regression test that ensures proper correction without inducing any new unacceptable behavior. A conditional release, which is justified based on a judgment that there is enough evidence to provide confidence that the validations will be met successfully, is based on production rate and usage in the needed scenarios. The conditional release supports the start of series production. A greater volume of production vehicles containing the system allows the validation targets to be met within a reasonable time after start of production. A projection can be made as to the date by which these targets will be met and the condition will be satisfied for the conditional release.

If there are no occurrences of an unwanted behavior, the conditions are satisfied, and a full release is recommended. Then the release documentation is updated to close out the condition for the safety case. If there is an unwanted behavior, then a field action might be necessary to contain it if warranted by the risk discovered. This is managed by the VM in cooperation with the supplier of the safety-related system. The root cause of the unwanted behavior needs to be ascertained and corrected. Depending on the severity and probability of the unwanted behavior, there may be a campaign to replace all the systems that have been released to the general public, or the root cause may be removed via a running change with no field action. As before, a rationale is provided for a practical regression test that ensures proper correction without inducing any new unacceptable behavior. This regression test is limited in scope to those use cases that are relevant to the anomaly and any other potential unwanted behavior based on engineering judgment. After the correction is verified and deployed, credit is reinstated toward completion of the validation target for unknown triggering events. When completed, a full release is recommended and is no longer conditional.

Integration of SOTIF and Functional Safety and Other Considerations

Background

The term *functional safety* is normally used with respect to potential hazards caused by faults in the system, while *SOTIF* is used with respect to potential hazards caused by the system without a fault. The ISO 26262 standard and the publicly available specification ISO PAS 21448 were written to address these differences. The two subjects could have been combined into ISO 26262, and this was discussed in the plenary meeting of the working group; there were arguments on both sides, since the safety of the systems being launched required both functional safety and SOTIF prior to series production and public sale. In the end, the proposal to combine SOTIF and functional safety into ISO 26262 was defeated by one vote.

Some enterprises deal with SOTIF and functional safety entirely independently. If SOTIF was included in the functional safety standard, it would be less acceptable in such organizations, and political difficulties might result. A functional safety organization is responsible for all three pillars of safety as they apply to functional safety, and the independence of these pillars is organizationally supported. This functional safety organization drafts the approved functional safety policy, perhaps referencing ISO 26262; audits the resulting process; independently assesses the work products; and ensures execution of the required tasks. One of the three alternative organizations discussed in Chapter 2 is employed, or a modification of these organizations is used.

Separately, a product safety organization may have a product safety policy that references regulations and safety standards other than functional safety, such as flammability or ISO PAS 21448; audits the resulting process; independently assesses the work products; and ensures execution of the required tasks. The responsibilities of this product safety organization are clear, and execution of a process to ensure the safety of products launched by the enterprise is effective.

If such an organizational split is made, all aspects of ISO 26262 are handled by the functional safety organization, and any other aspects of system safety are handled by the product safety organization. Using this separation, all aspects of system safety are addressed by two specialized groups, with any omissions due to the separation. In such an enterprise, it is fortunate that SOTIF was not added to ISO 26262. This content aligns with the responsibilities of these two organizations: SOTIF is in ISO PAS 21448, which avoids confusion.

Other enterprises have a system safety organization that handles both SOTIF and functional safety. There is only one organization structure to staff, and the safety experts are competent in both SOTIF and functional safety. The pillars of safety are not divided between SOTIF and functional safety but are combined. This allows the enterprise leadership to understand and support the combined organization with clarity. There is one system safety policy: it takes into account the requirements for functional safety and SOTIF and provides a basis for a combined process for the enterprise. The resulting process combines SOTIF and functional safety to ensure that the policy is enforced. This process is audited for system safety, including both SOTIF and functional safety. The engineers who attend these audits are expected to have elicited the requirements for functional safety and SOTIF and have evidence of compliance with these requirements. The process serves as a map to achieve these expectations. Each artifact that is assessed has content that addresses both SOTIF and functional safety. The assessment checklists used by the central system safety auditing and assessment organization when reviewing these artifacts include requirements for both SOTIF and functional safety when artifacts such as work products are reviewed.

Are there advantages or disadvantages to the two approaches of ensuring SOTIF and functional safety? Practitioners of each would probably say yes, based on the outcome of the vote when SOTIF was prevented from being added to ISO 26262. Much depends on the history of the enterprise and the safety organization. This history determines how much change would be required to maintain two separate organizations or to have one system safety organization.

If there was a regulatory or product safety organization first, and aspects of functional safety were handled separately in the development process prior to ISO 26262, or with an organization that relied on guidance directly from IEC 61508 but applied to the automotive lifecycle, then retaining this separation avoids organizational hurdles. Functional safety did not need to be handled by the legacy regulatory group, and regulations have generally not been the mandate of the functional safety group. There are specific experts in different regulatory aspects of safety as well as functional safety experts. These different experts have little reason to collaborate in their day-to-day activities. Tasks are executed independently, and interorganizational communication must avoid inconsistencies, which can be challenging.

If functional safety and SOTIF are combined, then some redundant tasks are avoided, which improves efficiency. The SOTIF process model and the functional safety process model have significant content in common. Consider the initial hazard analysis: this work product is needed for both functional safety and SOTIF, and there are many similarities in the hazard analyses. All the potential hazards that result from a functional failure, for systems that fall within the scope of ISO PAS 21448, may also be potentially induced by the system without a fault. Limitations of algorithms or sensors could lead to this outcome. Also, foreseeable misuse must be addressed by both SOTIF and functional safety analyses. Misuse of the system can affect controllability or even mechanically caused failures when addressing functional safety of the system. Foreseeable misuse is addressed by features added to the system, and other system modifications benefit both functional safety and SOTIF because they address how the system is controlled in case of failure and during normal operation. For example, consider monitoring the driver's hands on the wheel. Sensing hands-off and warning the driver may affect a lane-centering system's controllability when the lane boundaries are not detected due to sensor limitations, or control by the driver when the steering system loses assist due to a failure. Still, hazards induced by normal operation of the system, such as startling the operator, are normally considered by system safety and SOTIF, but are not considered when looking at hazards caused by system faults only. Startling the driver in case of a failure is considered by functional safety when evaluating controllability. However, this human-machine interface is SOTIF related and is considered in ISO PAS 21448.

Engineer training also benefits from the combination of SOTIF and functional safety. The training is pragmatic in nature and teaches engineers exactly how to accomplish all the SOTIF and functional safety requirements efficiently by combining them into existing work products. The process that includes these work products does not need to be altered except to include the additional SOTIF analyses and validation. Training recognizes that there is one engineering process, and everything must get done. This efficiency benefits customers with respect to resources as well as the cost competitiveness of the contract.

Analyses and Verification

Triggering event analyses, as discussed earlier in this chapter, is also specific to SOTIF. Triggering events are not discussed in ISO 26262 because failures of software and hardware are not induced by triggering events. The system is operating without a failure when initial triggering events are considered. Requirements are elicited by these analyses, as well. This is a systematic analysis, as previously discussed, and it helps provide confidence in the completeness of the requirements. Once the requirements are elicited and captured, they are included in the design and verified. The database that is maintained for functional safety requirements is extended to include SOTIF requirements. The verification can be made traceable to the SOTIF requirements to ensure its completeness. Then, corrective actions are completed to shore up any shortcomings, and validation ensues. These corrective actions include updating the requirements database and verification methods. Verification and validation methods vary between functional safety and SOTIF; both ISO 26262 and ISO PAS 21448 include guidance. There are tables for verification and validation methods in ISO 26262. SOTIF refers to the three areas mentioned previously in the chapter, which follow a prescribed flow.

A possible process flow combining SOTIF and functional safety is shown in Figure 3.1. Compared to the flow for a functional safety process only, some activities are expanded; others differ significantly between functional safety and SOTIF and are shown as performed separately. This flow assumes that an item definition describes the system's design and intent in enough detail to accomplish the analyses. This item definition or preliminary

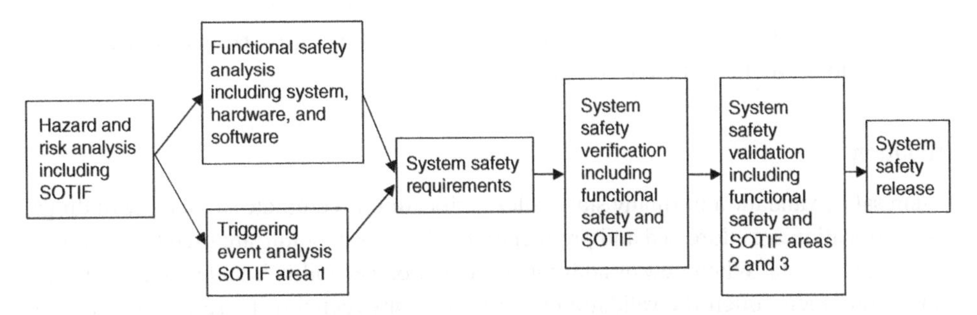

Figure 3.1 Example of a system safety process flow combining functional safety and SOTIF.

specification is needed for both functional safety and SOTIF. Complying with the item definition requirements in ISO 26262 should be sufficient for this purpose. The required item definition provides the system boundaries and functions with enough precision to support functional safety analysis and SOTIF analysis.

The hazard and risk analysis is performed, including all functional safety hazards and interactions of the system with the driver or other road users that may cause a hazard without a fault. This may be an increase of scope, when compared to a hazard and risk analysis for functional safety, depending upon the granularity of the functional safety hazard and risk analysis. The scenarios chosen for evaluating failures may have significant coverage of the use cases for SOTIF without a failure.

Then, the functional safety analyses of the design, such as a system failure mode and effect analysis, the single-point fault metric, and a HAZOP on the software architecture, are performed to elicit functional safety requirements. These analyses are separate from those required to elicit requirements or SOTIF.

The triggering event analysis is performed to elicit SOTIF requirements. As discussed previously, this analysis may have a very broad scope, depending upon the system's complexity and automation level. Requirements elicited by the analysis are included in the system safety requirements, and verification and validation plans are made, including all system safety requirements. The methods used for verification may be vastly different, but verification requirements such as traceability, a clear criterion for pass-fail determination, and documentation of the verification method are common. The requirements database links to SOTIF verification methods to facilitate the traceability of requirements.

Verification proceeds as normal in most engineering organizations. There needs to be a specification or database of requirements to be verified. A plan is made to perform the verification using specified methods. Some verification is accomplished by inspection, demonstration, and analyses. For example, the inclusion of a particular feature may be shown by demonstration only. Much of the verification relies on testing that has sufficient coverage, as described in ISO 26262. For example, the performance of a particular safety mechanism may be demonstrated by a fault-insertion test. SOTIF provides further guidance on verification, especially with respect to sensors. Requirements-based testing of sensors at specified limits of perception is particularly important for SOTIF verification because it provides information about potential sensor limitations that may be important when identifying triggering events. Evidence is compiled, as is the custom in most organizations, in test reports that are referenced in the requirements database. Traceability is provided to the particular test procedure used, the criteria employed, and the results. Actions are planned for any failures detected so that they are resolved. Any findings are corrected before system validation, to prevent redundant findings.

Validation

System safety validation normally starts after verification is complete so that any findings from verification are addressed and production methods and tooling are used for the product being validated. When new or additional production capability is added in the form of capital equipment, often the validation systems are constructed using run-off parts from the production equipment at the equipment supplier's location. This allows production intent systems to be used for validation at an earlier point in the program schedule. The intent of validation is to assure that no unreasonable risk is incurred by the general public due to noncompliance with safety requirements. This is true for functional safety as well as for SOTIF. Completeness of validation is critical to ensure this safety.

There are significant additions to validation for systems that contribute automated driving capabilities, and this validation is extensive because of these additions. They have a higher significance as automation levels increase. As discussed previously in this chapter, this includes validating the known scenarios in area 2 and the unknown scenarios in area 3. To maintain a launch schedule, significant time and resources may be needed to complete these SOTIF validation activities; that is why these are planned as soon as the requirements are known.

In a system safety organization where functional safety and SOTIF are combined, it may be highly advisable to employ an analyst to become familiar with the SOTIF concepts, who has sufficient creative, analytical, and mathematical skill to efficiently plan these validation activities. Preference should be given to employing a high-potential individual driven to derive and teach these methods to the safety and engineering staffs, as well as represent the enterprise in external forums in order to stay abreast of, and contribute to, the advancement of this technology. The analyst may, for example, use Bayesian statistics, employing the beliefs of the engineering experts in combination with testing to improve the efficiency of area 3 validation. Such techniques need to be employed with care such that their defense ethically may be assured, since the experts used may also be involved in the design. These plans for implementing Bayesian statistics need review of the assumptions, and significant resources are involved.

Evaluation of evidence from the steps just described leads to the system safety recommendation for product release. While the evaluation may be extensive, the recommendation must be clear and unambiguous with respect to safety sufficient for series production and public sale. This recommendation may be acceptance, conditional acceptance, or rejection for release; both ISO 26262 and ISO PAS 21448 have criteria for these three possibilities. A consistent rationale for this recommendation is needed within the enterprise. These criteria need a consistent interpretation by the system safety organization, which may be managed using guidance published within the enterprise to be referenced by those making the recommendation.

Acceptance indicates that the evidence of safety is sufficient. Usually, this judgment is made after reviewing the safety case, including the assessment reports for the work products and the audit reports. A minimum criterion may be that the evidence of safety is no less than for previous products of this type. Compliance with the standards may also be considered and is expected to be largely achieved.

Conditional acceptance indicates that there is no indication that the product poses an unacceptable risk, but further evidence is requested to confirm this. Perhaps an update to an analysis is required due to changes that are judged to not have imposed additional risk but need to be confirmed. Often a date is set for availability of this evidence.

Rejection indicates that there is evidence of an unreasonable risk, or the evidence available is not sufficient for conditional acceptance. The product may be rejected because the evidence is less than for previous products. Perhaps a needed analysis is missing, and released products include it.

4

Safety Audits and Assessments

Background

Chapter 2 discussed the pillars of a safety organization. The first pillar is the enterprise system safety policy, the second pillar is independent audit and assessment, and the third pillar is execution of the enterprise system safety policy using the process it supports. These three pillars are independent of one another, although they do interact. The independent audit and assessment activities consist of audits of the system safety process and assessment of the system safety work products. These work products are executed independently of the audit and assessment activities, within the enterprise product lines as part of the specific programs to launch automotive products for series production and public sale.

Chapter 2 discussed several approaches to accomplish this within an enterprise that produces automotive products, either as the supplier or the vehicle manufacturer (VM). Different organizational approaches can be used to execute the work products and also maintain the independence of auditing and assessment as well as the enterprise system safety process. We have discussed the advantages and disadvantages of each approach. In every case, audits and assessments produce value by ensuring that the safety risk of releasing the product is not unreasonable, and that evidence supports an argument of diligence. The independence of the auditing and assessment activity is critical to providing an unbiased, credible argument of safety. Such independence is required by accepted standards, such as ISO 26262. Compliance with accepted standards strengthens this argument, and audits and assessments support compliance.

Audits

The audits discussed here are performed on the engineering project during or after development and before a recommendation for release is determined. It is important to perform these audits during the project in order to enable any corrections or improvements that are needed in order achieve a favorable recommendation for release. The purpose of the audits is to confirm that the correct steps have been taken, that the timing is appropriate for release, and whether any additional actions need to be taken based on the findings. Such findings may involve timely completion of work products, and capture of requirements

Automotive System Safety: Critical Considerations for Engineering and Effective Management,
First Edition. Joseph D. Miller.

such as assumptions for the use of any safety elements out of context – for example, a microprocessor or software operating system. Findings may also address required completion of safety analyses before the start of verification activities in order to have confidence that the requirements are complete.

The audit also provides an opportunity for the product staff to raise any additional concerns to someone independent of their direct supervision without fear of repercussions. These concerns are noted by the auditor, who may address them outside the audit with the product line staff or the executive staff of the enterprise. These may also be noted as findings of the audit or in the audit report.

In addition, the audit report provides senior executives responsible for the product with the status of the achievement of system safety, including any escalations that the auditor determines merit special attention, such as staff concerns with respect to safety. Because the audits are conducted in advance of the program's major milestones, information from the audit reports may prepare the executives for reviews at those milestones. This allows executives to follow up on the actions and escalations and otherwise support the achievement of system safety through the allocation of resources. Because the audit reports to engineering executives and program management precede the senior executive reviews of product line programs at major milestones, there is time to improve the status that is expected to be reviewed with senior executives so that it is more favorable. Experience supports that engineering executives will use this opportunity of time to improve the safety status to compete with their peers in other product lines. It helps the executive differentiate the product. Thus the audit is pragmatic.

During the audit, as findings are discussed, an auditor may facilitate resolution by explaining how similar issues have been resolved on other programs. The auditor may audit many programs in diverse product lines in multiple cultures and with many different statuses. Nevertheless, it is not unusual for similar types of problems to exist in such audits and be resolved. The broader the auditor's personal experience in operations and engineering, as well as safety management and assessment, the greater the breadth and depth of these explanations. In general, such explanations are welcomed, particularly when available resources for resolution are provided or referenced, such as an in-house or external contractor or supplier, and a colleague's contact is provided as a practical reference, even if the colleague is in another product line. This can be a valuable part of the audit: expectations of assistance help to overcome participants' inertia about coming to the audit when they are working on the program under intense time pressure, as is common on automotive launch programs.

The audit should be organized to minimize disruption to the project. There should be an expectation that the audit will be useful and to the point, and will follow a consistent and well-understood agenda. This can be accomplished by minimizing the time of the audit and maximizing its efficiency. For example, sometimes an audit is scheduled for the end of the project. This is not necessarily efficient or of the greatest value to the program team. The auditor examines the available evidence, which is from the entire duration of the project and is time-consuming to study. Safety requirements are reviewed in detail for traceability, evidence of verification, and evidence of compliance to methods of standards. At the end of the project, there is little or no time to improve these – the verification phase of the program has already passed. Teaching opportunities are minimal, and status reporting may be later than is useful to support improvement by the program team and executives. Action by product line executives to improve the status of the program may not be plausible. The audit may last several days, during which many people are involved but receiving limited benefit. This is not efficient.

Alternatively, several shorter audits can be scheduled at different points in the project's progression. Audit dates are selected to maximize utility and maintain an on-time status: for example, an audit after the project kick-off in the concept phase, another audit during the design phase prior to verification, a third audit during or after verification, and a fourth near the end of validation and project release for production. The objectives of these audits are selected for relevance at each of these phases of the program. For example, the audit after kick-off focuses on the concept phase, changes from the baseline program, the safety impact analysis of these changes, and tailoring of the current safety lifecycle.

The audits can be scheduled so that the people involved with different topics are present while their part is being discussed. This can be accomplished by organizing each audit in accordance with domains of expertise. For example, all the software work products and analyses can be organized for one time period on the agenda. The same organization can be used for electrical hardware tasks, such as architectural metrics and fail operational redundancy requirements, if relevant.

The product line executive receives the audit report information in a timely manner, and this increases the usefulness and value of the information. Priority information is summarized, and expected actions by the executive should be so clear, concise, and practical that no questions are required and the executive can issue directions at once while viewing the brief report or summary email.

Such an audit of a complex project can be completed in two hours, with sufficient planning. To accomplish this, the participants must be prepared, the current status of previous audit actions needs to be available, and participants must show up on time or have a delegate attend on time. For a project with enough valid carryover content from another project with a safety case, the audit may take less than hour. The participants must be just as well prepared, but the scope is smaller. Such an audit requires significant preparation of the project team by an excellent and well-respected safety manager. The safety manager helps the project team members to understand precisely what is expected and to prepare. For example, the product team can be so completely prepared for an audit that they anticipate the auditor's questions and have slides prepared with data and calculations to prove that previous issues were resolved mathematically. Experience confirms that this is possible, if preparation is well disciplined.

Audit Format

A standard format helps make an efficient periodic audit possible. This allows the responsible safety manager, assessor, and project team to prepare. An example of what is included in such an outline follows:

1) Description by the program manager or project manager of what is included in the program, and any changes in scope or significant dates since the last audit.

   This may consist of, for example, a set of slides previously used for a briefing, along with a schedule.

2) Follow-up on actions from the last audit. This is prepared in advance and recorded, for example, in a standard template used across the enterprise.
3) Additional actions due to #1. Include a description, responsible person, assigned date, due date, and status (e.g. assigned). These are prepared during the audit and recorded, for example, in a standard template used across the enterprise.
4) Verification that due dates for work products correspond to the dates from #1. It is not uncommon for project dates to change in an automotive program while the program is being managed.
5) Review of the status of work products that are due. Include a review or walk-through of significant work products, such as the impact analysis if significant carryover credit is taken (assumes a detailed review of work products is performed by an independent assessor as required). Record any actions required. A standard template used across the enterprise can be used; it is convenient if it is sortable by due date and other attributes such as completion status.
6) Review of requirements traceability from elicitation of system requirements to hardware and software requirements to verification, including methods. This includes all three sources of requirements (discussed shortly in this chapter). Record any actions required. Direct review and auditing of a requirements database and any information traceable to it is an efficient way to audit this. Auditing of requirements is a priority in each audit, because work products do not provide this information.
7) Check if there are other issues that anyone wishes to raise independently. This can be as simple as an open invitation by the auditor prior to reviewing the summary audit notes.
8) Review the audit notes (taken by safety manager or assessor) that will be presented to the product line executive, to confirm agreement. Note the metric status reported. While the audit notes may or may not present a favorable status, as reflected in the safety metric, it is important that the project team understand and concur that it is accurate based on the enterprise safety policy or process criteria.
9) Close the audit.
10) Later, send the report, with a summary cover, to everyone invited and the appropriate executives, noting attendance, absences, and substitutes as well as preparedness and participation. This can be a brief report; nevertheless, the transparency it provides promotes credibility and trust in the auditing process.

Such an audit, as outlined here, can move very quickly and efficiently. The steps are ordered so that they proceed logically. Only relevant information is reviewed with the people involved. Continuous improvement in the execution of the audit is supported. It also allows the attendees to receive credit for their attendance and readiness, which are made visible to the product line executives. This helps preclude non-participation due to project timing pressures, because invitees' self-interest provides motivation for their attendance. Minimal time can be allocated with maximum benefit for all involved. Everyone can see and agree on the safety status for all phases of the program, which helps prepare for other reviews. Experience indicates that this is effective.

The approach outlined here may appear Machiavellian [7] to some, as if the auditor is seeking a princely dominion over the product team to capture system safety for the enterprise. "Is it better to loved or feared?" you may ask. The answer may be, "Both." The audit is not about the auditor but is about efficiently determining and reporting the status of the program's safety process in order to facilitate compliance with the process. It is ideally accomplished while expending minimum time and resources on this determination and reporting.

An auditor following this approach secures maximum benefit for the product team, because the audit enables the team to gain positive visibility for themselves for excellent performance in ensuring system safety for the product being launched by the enterprise. Executive leadership expects this excellent performance to add value by minimizing the risk to the enterprise of a safety issue in the field that may be traced to insufficient diligence with respect to the safety process.

The auditor may also bring solutions to the product team that save them time and budget. These may come from synergy with other programs and product lines enabled by communication across product lines by the auditor. When help is needed, the independent auditor identifies it independently so the request does not appear self-serving to the product team. The need and the action required are communicated actionably to the executive with the required authority. All of this is appreciated by both the executives and the product teams, and the audit may seem to add more value than the resources it consumes.

Nevertheless, there may be motivation to delay or deter an audit report that is not completely favorable by attempting to intimidate the auditor. This behavior is exhibited by program or product line personnel as a way of pushing back against a report that appears unfavorable. Even the appearance of an attempt to intimidate or threaten the auditor must not be tolerated. Any such attempt must be swiftly, emphatically, and openly corrected. There must be no indication that intimidation will ever succeed in deterring reporting of the safety status for a program leading to series production and public sale. *Openly* does not suggest that this be done outside the enterprise or on a public stage: it is not the intent to expose the enterprise to public criticism. In fact, it is just the opposite, because by internally preventing attempted intimidation, criticism can be avoided. Still, anything internally written could be made public in an unfavorable manner. Internal written communications must be prepared with care and in accordance with advice of counsel that the enterprise provides to employees. In fact, the use of attorney-client privileged communication requesting further advice of counsel may be warranted for the corrective action. This provides some protection for the privacy of the communication and an opportunity to receive the advice of counsel concerning the attempted intimidation. Still, anyone who witnessed the attempted intimidation must also witness the correction, in order to preserve the enterprise's safety culture. They could be included on the attorney-client privileged communication so that they are informed of the resolution of the intimidation attempt and also receive any advice that counsel provides.

In addition, if the auditor makes a mistake, the auditor must acknowledge it. This must also be done swiftly and emphatically. Integrity of the auditor must be ensured, because the auditor leads by example.

Use of External Auditors

Project audits are sometimes performed by external organizations. There is not a single safety organization or accreditation endorsed by safety standards such as ISO 26262, and no certification is recognized by the automotive safety standards. Such accreditations and certifications are provided by other organizations, such as consultancies, which also provide audits for programs under contract if requested. Such audits may be periodic or at the end of a project, as discussed previously. The determination of what audit services are provided is made at the time the engagement contract is agreed to.

External auditors are considered independent because they report to an external organization and not to the project's release authority. This satisfies the independence requirements of ISO 26262, for example. Talented, experienced external auditors can bring experience to the audit that is not available internally. The auditing organization may have provided audits for other automotive companies as well as other industries. This can provide a fresh, broader outlook to the evaluation of requirements traceability as well as potential solutions to problems encountered. Techniques to ensure or evaluate traceability can be shared, as well as resources to implement the resolution of problems. A customer may specify external audits and require a specific auditing entity, or the customer may perform the audit.

External audits managed by each product group can undermine the assurance of consistent audit practices across the enterprise. Each product group could choose a different auditor, for example, if sourcing is at their discretion. Such sourcing decisions can have different criteria: one product line could be audited in a manner that would have benefited another product line if it had been consistently applied. Such sourcing may not promote consistency because consistency with other product lines is not a sourcing criterion. Cost and availability may have higher priorities. In addition, there could be an argument against independence if the release authority is also the contract manager for the auditor sourced. Then the project being audited pays money to the auditing entity, and financial independence might be challenged.

The benefits of external audits can be achieved while avoiding the disadvantages, if the external auditors are contracted by the central organization. The central organization includes consistency in audits from product line to product line as a criterion for selection of the external auditor. The auditing procedure adopted by the central organization is executed by a qualified external auditor who is observed by an internal auditor and vice versa to ensure consistency with internal auditors, just as would be done if an internal auditor were added to the auditing staff.

The external auditor is independently selected, and this selection matches the auditor's talent and experience to the projects to be audited. For example, an auditor who has particularly extensive knowledge of and experience with the development of safety-related software can be matched to a software product hosted on an external hardware platform that supports more than one vehicle function.

Feedback from audits conducted by the external auditors can be used for continuous improvement by the central organization. External auditors provide feedback concerning the need to address potential systemic issues of the product line or enterprise from an external perspective, and product groups provide feedback about the auditing styles of external auditors that they find particularly beneficial. Similarly, poorly performing external auditors can be screened from other product lines; their behavior can be addressed, or they may not be engaged in the future. It is assumed that the central auditing and assessing organization is not directly funded by the product lines, because such funding could adversely affect the optics of independence. Audit resourcing should not directly depend upon the entities audited. Independence is ensured.

Assessments

System Safety Assessment

The final functional safety assessment is discussed in ISO 26262. During this assessment, the overall safety case is evaluated to determine whether it is accepted, conditionally accepted, or not accepted as being sufficient for product release for series production and public sale. Also discussed are assessments of work products; ISO 26262 has requirements for independence to review these work products, as well.

These two types of assessments – the final functional safety assessment and the assessment of work products – can be combined if all the evidence of system safety with reference to the work products is collected in a final safety assessment by the safety manager and that document is independently assessed by a system safety assessor. Such a document is a work product included in the system safety process and also includes the evidence of the safety of the intended function (SOTIF), for example, in accordance with ISO PAS 21448. The applicable tasks required for SOTIF are assessed as well. In this manner, efficiency is gained by enterprises that employ a single system safety process that includes both SOTIF and functional safety. This efficiency is enjoyed by both the product lines executing the process as well as the central safety organization. This procedure is deployable across an enterprise.

If functional safety and product safety are separated in the organization, this procedure can be deployed separately by each organization. Similar efficiencies can be realized by the separate organizations. The procedure just described replaces a final safety assessment described in ISO 26262 and the report prepared solely by the assessor. This removes a program delay required for the ISO 26262 final safety assessment. The procedure described is scalable.

Work Product Assessment

Assessment of work products is different from an audit or a design review. An audit is performed to review process compliance and examines work products for evidence of compliance to the process. A design review is concerned with technical evaluation of the suitability of the design. The system safety assessment of work products confirms that the requirements for the work product are satisfied and that the work product is suitable for inclusion as evidence in the safety case. The requirements for these work products are derived from referenced safety standards, as well as internal guidance.

A work product for the process may pull requirements from several work products of a safety standard to gain efficiency, reduce the number of work products, and align with other standards, such as quality standards. Procedural attributes are checked, such as correct references; reference correctness is of high importance because of the need to be able to accurately retrieve evidence for the safety case. Correctness of statements is checked, because inconsistent or questionable rationales can adversely affect the safety case. For example, in an impact analysis for a product, an argument that a work product may be reused is checked to see whether the work product is retrievable and confirmed, as well as the rationale for its reuse. If design changes have an impact on this rationale, then the work product may need to be reworked rather than completely reused.

These assessments are more extensive than an audit; the audit checks that these assessments have occurred. The assessments are much more granular and focused on the individual work products, not the entire execution of the safety process on a program. They are different from a design review in that they are not intended to verify the correctness of the design itself. Such a technical review is expected to be completed and approved prior to the work product being submitted for assessment. Lack of evidence of such a review is a sufficient reason to return the work product to the author without being assessed, because such an assessment is invalid if the technical content is not correct. Such assessments can be resource intensive, and efficiencies can be sought.

Does assessment of work products need to be performed internally, or can it be outsourced? Consultancies can be sourced under contract to perform detailed assessments. The capability to check the correctness of internal referenced document would need to be provided to the external assessor, or this would need to be separately accomplished internally.

To some extent, the same principles discussed concerning audits apply to assessment of work products. While the purpose of performing an audit is different from the purpose of performing an assessment of a work product, the independence considerations are alike. There are advantages to gaining an external interpretation of fulfillment of work product requirements, but varying interpretations can also lead to inconsistencies and confusion. For example, suppose a hardware engineer has performed single-point fault metric (SPFM) analyses for an airbag diagnostic module for many projects, and this work project has always been assessed by an internal assessor to I3 independence. An external assessor assesses the next SPFM and does not agree with the coverage claimed, perhaps by a redundant path and comparison. Similar reviews by customers have agreed with the previous assessments. Now the different assessment interpretation must be resolved.

Where independent assessment is needed, such as the impact analysis, hiring of an external consultant by the product line raises the same challenges as for an independent audit. The impact analysis is key to determining the resources required to generate the safety case because it determines the necessity for generating new or reworked work products. The financial motivation and the sourcing of the assessment may be questioned regarding independence on this account.

Resolution of these disadvantages can be mitigated by managing any supplement to the internal assessments with external assessments by the central organization. Then, even if the same consultant is hired to assess the impact analysis, the question of independence is removed. Again, the central organization should be financially independent of the product lines. Consistency can be ensured centrally, and independence is guaranteed.

Unlike audits, portions of the assessment can be performed remotely, perhaps in a lower-cost country, by an internal unit of the enterprise or an outsourced enterprise. With proper access to the enterprise's infrastructure, references can be checked remotely and a preliminary assessment can be done using a checklist. Sufficient skill should be ensured, and the remote assessments should follow a process approved by the central system safety auditing and assessment organization. This helps ensure uniformity and completeness of these assessments, which are provided to multiple internal assessors as a baseline for their final assessment.

The remote resources can keep overall metrics on the causes of rejections for nonconformity. Criteria for these are refined through continuous improvement toward consensus criteria for rejection and can be used to improve training for the enterprise: they serve as a pareto of topics to be emphasized and explained by the trainers. Then an assessor with domain knowledge of the product can review the work product and finalize the assessment. Feedback about any changes is shared with the offshore partner, providing an opportunity for improvement. Any corrections can be discussed with the author in person. The personal one-on-one discussion is beneficial, and this is a risk-free opportunity for coaching and mentoring; future assessors and safety leaders may result. This assessment and mentoring are scalable and sustainable.

5

Safety Culture

Background

ISO 26262 discusses *safety culture* as being the enduring values, attitudes, motivations, and knowledge of an organization in which safety is prioritized over competing goals in decisions and behaviors. This is a clear view that implies a bias for action to ensure safety whenever a decision is made in the automotive industry about what action to take next. Cost, schedule, and political urgency must not be valued above ensuring that the general public bears no unreasonable risk due to an enterprise's actions or omissions.

While this is with respect to automotive functional safety, no new definition has been proposed for safety of the intended function (SOTIF) by ISO PAS 21448. *Safety* is still the absence of unreasonable risk when used in the context of SOTIF. Therefore, a *safety culture* implies a bias to ensure no unreasonable risk. Thus, this concept of safety culture seems broadly applicable to system safety. The purpose of system safety is to ensure the safety of the product when operating without failure as well as in the presence of failure. The organization is motivated to achieve safety, and the safety culture provides empowerment.

ISO 26262 discusses positive and negative indicators of a safety culture that show whether the persons responsible for achieving or maintaining functional safety have dedication and integrity, as well as the persons performing or supporting safety activities in the organization. It is difficult or impossible to measure a culture directly. Still, a safety culture is important to establishing and maintaining a process that supports achieving no unreasonable risk for automotive products provided to the general public. Also, safety thinking throughout the organization encourages questioning and excellence and avoids complacency. Every change to a product or requirement is examined for its safety implications. Whenever an activity is considered for tailoring, the enterprise also considers whether the general public would be unknowingly assuming a greater risk as a result. The organization fosters taking responsibility and self-regulation.

Characteristics of a Safety Culture

Possible positive and negative characteristics of safety culture are listed in ISO 26262. Positive indicators include the following:

Automotive System Safety: Critical Considerations for Engineering and Effective Management,
First Edition. Joseph D. Miller.

- Accountability for decisions is traceable. This implies, for example, that an identifiable person is responsible for the actions taken and not taken. This includes both design and planning.
- The highest priority is safety. For example, the rationale that there is not time to verify all safety requirements in a traceable manner is not acceptable.
  The reward system favors safety over cost and schedule and penalizes the opposite.
  For example, using metrics, safety goals are set for each product line. At-risk compensation of the executives in each product line is weighted most heavily for safety goals. This is cascaded throughout the product line and ensures a congruent safety motivation.
- Checks and balances are assured by processes, including proper independence. Evidence of implementation of the three pillars of safety can ensure this.
- Issues are discovered and resolved early, including management attention before field problems occur. Periodic audits, assessments, and metrics reviews by senior enterprise executives can provide evidence of this.
- Adequate qualified resources are provided to execute safety tasks early. Ownership of the system safety tasks by program management can provide for this resourcing, provided that failure to adequately plan and resource the safety tasks is periodically reviewed by senior executives.

Negative indicators are juxtaposed to describe the lack of these positive indicators. For example, vague accountability for decisions, schedule, and budget prioritized above safety are negative indicators.

A system safety culture implies a commitment to elicit a complete set of system safety requirements and a commitment to comply with these requirements. Providing evidence of this achievement is the fundamental purpose of creating a safety case. While it is expected that agreeing with this commitment is important, embedding the commitment into an organization as an enduring value may require actions that demonstrate executive support and thus promote empowerment to perform these tasks. Such actions may come from the central safety organization, the project safety manager, enterprise leadership, and even customers. The central safety organization performs auditing, assessment, teaching, mentoring, and technical safety backups to facilitate successful execution by the product teams. The product line safety manager keeps the day-to-day system safety process on track through task execution, coordination, and communication with the customer's or supplier's safety manager. The customer supports these shared system safety priorities. These contributions are summarized in Figure 5.1. All contribute to the organizational values and play important roles.

Central Safety Organization

The central system safety organization is expected to have deep domain knowledge of system safety. This knowledge may come from not only personal study and training, but also from many years of project participation in engineering and operations as well as in system safety. System safety experience can include a variety of safety analyses and also experiencing the complete safety lifecycle for automotive products, which provides a fundamental understanding of the nuanced causes and effects of actions in each lifecycle phase with respect to the other phases.

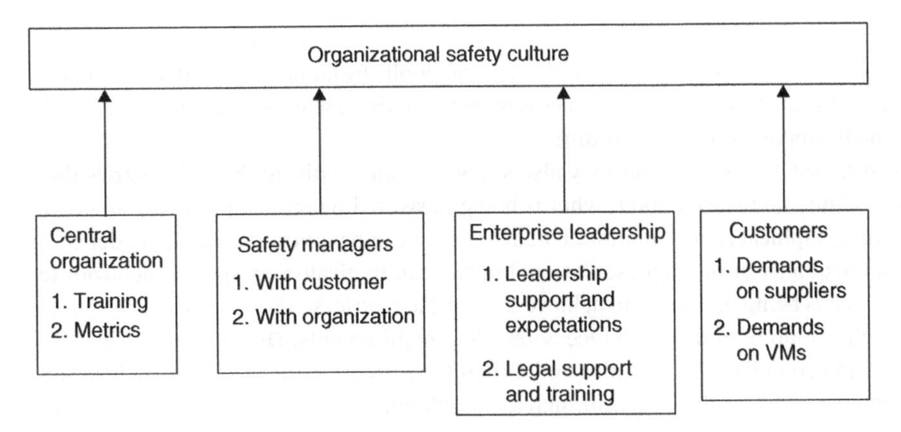

Figure 5.1 Factors influencing an organizational safety culture.

The central organization is expected to know what values should endure in the safety culture. There is a deep and passionate belief in the "true north" of systematically eliciting a complete set of safety requirements early enough to ensure compliance and establishing evidence systematically that compliance has been achieved so that no one in the general public is exposed to unreasonable risk from the product. These values should motivate the central organization, which is why the central organization provides assessments and audits of the system safety process and artifacts. These actions are driven by a belief in providing safe products to the general public. Since every action must have a motivation, the actions of the central organization implicitly demonstrate the belief system and enduring values of the underlying system safety culture. Meticulous checking, coaching, and formal teaching show an attitude of prioritizing system safety, and reports and metrics measure the effectiveness of these actions.

Formal training by the safety organization helps increase the knowledge of the organization and ensure that the values endure. Formal training is made mandatory by the product line leadership as well as by enterprise leadership. When training enterprise executives, management, and engineering in an enterprise that releases safety-related automotive products to the general public, emphasis is placed on the ethical behavior that is expected from these roles. This responsibility for ethical behavior is taught not only from a legal point of view but also from a regulatory point of view, as well as from the perspective of cost and business reputation. This ethical behavior may affect whether the organization is eligible for new safety-related business awards.

Leadership is taught that when assessing the risk of releasing a product by a deadline, it is not okay to expect the general public to accept the risk for safety-related tasks that were not completed on time. For example, comments from a manager such as "I have high confidence and am willing to take the risk" are not acceptable. This risk is being imposed on the general public, not the manager, without their consent. Training reinforces the priority of system safety in the enterprise by consolidating and clearly communicating the culture demonstrated by the central system safety auditing and assessment organization.

Training also teaches what is expected in terms of eliciting requirements through analyses and verification. Its importance can be internalized more easily by considering the risk imposed on the general public when this is not done. No one in the enterprise would deliberately impose an unreasonable risk on the general public by launching an unsafe product. Connecting this to the system safety process is key: then those who are taught are self-motivated, which supports the safety culture.

The central safety organization may also support a safety culture through metrics that provide visibility and clarity about what is being measured by reducing it to a numerical value with a graphical representation. Understanding can be quick and intuitive, depending upon the quality of this representation. Metrics can motivate very specific behavior to improve what is being measured to achieve a favorable metric, so they should be chosen to motivate elicitation of and compliance with safety requirements. Directly measuring this elicitation and compliance is complex and difficult at a product line or enterprise level; an abstraction that is indicative of them, such as completing the safety analyses on time with assessment acceptance, is sufficient.

Safety culture is considered a property of the organization, so a metric can motivate this by measuring whether the system safety process is applied on every product early enough to be successful. This may require multiple points of measurement. Program milestones are useful to institutionalize these points of measurement: for example, an early milestone of a successfully assessed work product can be measured, to indicate whether the process is being applied on every program. Another milestone can measure whether all the analyses necessary to elicit derived safety requirements have been completed in time to be verified. Such a milestone agglomerates all the work products that represent safety analyses and takes into account any credit for reuse of prior safety analyses. (To some extent, this may encourage reuse and have incidental financial benefits to the enterprise.) Other milestones are chosen to measure intermediate progress, as discussed in Chapter 4 with reference to safety audits. These may include, for example, verification and validation as well as completion and acceptance of the safety case. The audits are then supported by the metrics. Senior management reinforcement helps process compliance. A favorable safety culture emerges.

Safety Managers

The safety manager is embedded in the product development group, regardless of the enterprise organization. In Chapter 2, Figure 2.3, the safety manager role is assumed by the domain manager, such as the system engineering manager or electrical hardware engineering manager. This role presents a unique opportunity for fostering the organization's safety culture by contributing to and facilitating the execution of the system safety process on each launch program. The safety manager can lead by example, encouraging development colleagues to appreciate the value of this dedicated work.

If metrics are provided by the central system safety auditing and assessment group, then the safety manager is the point of contact to communicate the system safety process status for the program. Therefore, if the safety manager's leadership and help result in improving the system safety metrics, colleagues and product line leadership are supported and rewarded. Support of the development safety tasks is appreciated, and this dedication is reinforced.

Joint Development

This is especially true when the safety manager of a vehicle manufacturer (VM) and the safety manager of an automotive supplier engage in a joint development. In this case, both safety managers are expected to serve as the point of contact for safety communication. The priority of safety tasks is jointly reinforced in both organizations; this can be ensured by making communications transparent to the entire product line organization. The VM's personnel are made aware of what is needed by the supplier and why, as well as any impact on the schedule that may result. Likewise, the supplier's personnel are aware of what is needed by the VM to achieve the required program milestones. There is an urgency to coordinate the safety requirements in order to support a design in time for the B samples (design intent systems and vehicles) used for verification. Any incompleteness of the safety requirements by verification may result in incomplete or delayed verification. This is unacceptable if the B samples are intended for use on public roads, including fleet testing of SOTIF-related products. Such milestones and strategies are jointly planned, and a mutual safety culture develops.

Each safety manager carries these needs for completeness of requirements elicitation back to their organization. Required safety analyses are executed by members of the various domains on the program, such as the software safety analysis and hardware safety analysis. Now the safety tasks to achieve the launch milestones are executed as a team. Each team member understands what is required, and the value of the tasks is apparent in every review. Thus the safety culture deepens.

Enterprise Leadership

Responsibility for every product on a launch schedule is ultimately owned by the executives of the enterprise. These executives bear the cost, schedule, performance, and safety responsibility for every product put into series production and sold to the general public. Any risk placed on the general public by nonperformance of safety tasks that should have been known or executed will ultimately be a risk to the enterprise and perhaps personally to the responsible executive.

Formal training provided by the central system safety auditing and assessment organization as well as briefings provided at reviews, such as metric reviews, reinforces executives' awareness of this responsibility. They must decide what to do to prevent nonperformance. The system safety auditor may provide answers that support executive action in audit reports for programs within an executive's authority. When the executive takes this recommended action, awareness of the executive's safety responsibility is promoted. This supports a positive safety culture, because others are motivated to take action in the areas of their safety responsibility. The priority placed on safety goals can cascade to the goals at each level and function of the organization. All are involved in achieving these goals together, which empowers management assignment of resources. Thus executives can foster safety culture.

Liability

Questions of liability often revolve around what is necessary to demonstrate diligence with respect to safety. This may include determining what the *state of the art* is with respect to safety at the time a product is launched. This is true of functional safety as well as SOTIF.

Some of the motivation for standards relates to liability. For functional safety, evidence of compliance with ISO 26262 is intended to show diligence with respect to the methods and measures used to develop and manufacture a safe product. However, this evidence may require interpretation and may not be sufficient to demonstrate achievement of system safety diligence. Diligence may also be supported by an argument of compliance with ISO PAS 21448, although this PAS also requires interpretation and this interpretation may evolve with use.

At many safety conferences, lawyers present papers on subjects related to liability. There is some consensus in these presentations, though they are general and not sufficient to support liability concerns arising in everyday work regarding complying with an enterprise's system safety policy and process. However, advice and counsel from the enterprise's legal department is precious to the safety culture of the organization. The vacuum in an organization that is missing such counsel may be filled by assumptions or other sources of information that is not endorsed by the enterprise's legal counsel.

Attorneys within the organization who have litigation experience with automotive products have special credibility with respect to liability and can be highly influential in promoting a positive enterprise safety culture. They are talented and obtain organizational prominence that further adds to their credibility with product line personnel.

The domain of the legal department is law, not safety standards or system safety, so system safety training that includes the legal departments should not be overlooked. To obtain well-informed legal counsel with respect to safety process and any associated liability, the legal department should have overview training with enough granularity concerning process and intent to enable them to provide informed legal counsel.

The safety process generates work products and arguments that are maintained as evidence of compliance, much like engineering drawings and failure modes and effects analyses (FMEAs). Such evidence or the lack thereof may be the subject of counsel from the law department. System safety process metrics measure attributes of process compliance, and the definitions of these attributes, along with reviews and evidence, prompt action that supports evidence of diligence and elicits useful counsel from the legal department with respect to metric definitions.

The legal department's vocal, active support of the system safety process, the evidence it produces, and system safety metrics is critical to promoting a safety culture that embraces transparency. This support should be made visible to the entire enterprise organization on a periodic basis, such as during system safety workshops or periodic meetings of the combined safety organization. In particular, legal training for safety, engineering, and program management personnel that includes liability topics related to safety is very helpful, because it promotes an understanding of the legal counsel it provides. Some legal training can be provided by outside counsel; while it may be accurate, it may not have the same impact as internally supported training. Having this training provided by internal counsel adds credibility and encourages questions – even if the material and documentation are identical to training provided by outside counsel. Online training, prior to training by counsel, may further support such open discussion. This online training may provide a basic understanding as well as raise some additional questions. Then the attendees can prepare for the legal training by counsel. This preparation promotes deeper personal understanding and is internalized.

Customers

As discussed extensively in Chapter 1, consumer expectations provide a compelling motivation for VMs. Potential consumer expectations must be satisfied in all areas sufficiently to lead to the purchase of the vehicles being manufactured. These include expectations related to system safety, because requirements that relate to the consumer's safety are elicited and compliance is ensured. All system safety requirements are driven by a need to satisfy the consumer's expectation of safety. As an awareness of this connection is made widespread throughout the enterprise, a safety culture is fostered. This starts with understanding potential hazards and providing training that promotes the understanding that all system safety requirements relate to avoiding those hazards. If the VM has a passion to provide the highest possible level of safety to the consumer, derived from both ethical and competitive considerations, and the VM strives to delight the consumer, this will drive the safety culture.

Similarly, this passion cascades to automotive suppliers and throughout the automotive supply chain. It is fostered by connections and sharing values. Each automotive supplier strives to satisfy customer requirements, including safety requirements. Suppliers develop a process to ensure that customer satisfaction is sustainable, scalable, and maintained. As the supplier demonstrates competence and dedication to eliciting a complete set of safety requirements and complying with those requirements, the supplier enjoys greater confidence from the customer. The supplier strives to earn this confidence and sustain a favorable relationship with the customer.

There are compelling financial as well as program reasons for the supplier to achieve this confidence from the VM. Future business may depend on it. In addition, execution of the program and reviews is more efficient, and the customer's recognition of competence helps promote a safety culture within the supplier. Program personnel sense this confidence as appreciation of their efforts to satisfy the customer in the area of safety. The customer may make the supplier's executives aware of this recognition, and the executives in turn may recognize the system safety achievements of the program team, thus further promoting a positive safety culture. In this way, customers can help prioritize safety.

Safety Culture vs. Organization

The different organizations discussed may have an influence on effectively promoting a safety culture. Every potential organizational structure is a trade-off of several factors, and the final organizational structure represents the trade-offs that were accepted. The structure itself can influence the perceived value of safety in the enterprise. Part of this perception is based on the position of system safety management and experts in the hierarchy. If the entire safety organization reports solid line to enterprise leadership, as in organization 3 (Chapter 2, Section 2.3), then a higher value is perceived. Trade-offs among direct reports must be made, as well as in staffing levels of the system safety organization; this adds credibility to processes supported by the safety organization. Executive visibility provides favorable optics that indicate a higher priority of system safety within the enterprise. The reward system is consistent with the enterprise's values, and this supports the safety culture.

Consider the organizational structure of alternative 1 in Chapter 1, where independence is maintained among the three pillars of safety. In this alternative, the central system safety organization performs audits and assessments while reporting to enterprise leadership. Enterprise leadership supports the auditing and assessment function by providing a system safety policy for the enterprise; and auditing and assessing the system safety process and artifacts ensures compliance with this policy. The safety manager reports to product line leadership and manages the system safety experts in the product line. This provides direct support for the safety experts as well as analyses performed by the other engineering disciplines. Having this reporting structure within the product line fosters a safety culture: the support is visible, effective, and focused on the needs of the product line.

System safety is perceived as having an equivalent status to system engineering or software engineering in a single-product organization, or a status equivalent to the leader of a product in in a multiple-product-line organization. Equal treatment of system safety priorities is expected on projects and in engineering reviews. This improves the culture of including system safety in decision-making. In a matrix organization, the system safety manager is a resource manager. System safety priorities have the same priority and consideration in product line engineering decisions. The reward system is consistent with the enterprise's values, and this supports the safety culture.

Organizational alternative 2 in Chapter 1 may not be as favorable for supporting a safety culture. Independence is maintained among the three pillars of safety, but product line leadership assigns system safety to one of the functional managers, such as electronics or system engineering, and this manager acquires or hires system safety experts. As in alternative 1, enterprise leadership support the auditing and assessment function by providing a system safety policy for the enterprise; and auditing and assessing the system safety process and artifacts ensures compliance with this policy.

The status of system safety in the organizational hierarchy is diminished, as compared to each of the other alternatives. System safety experts may be contracted and not direct employees of the product line. These experts are assigned to support execution of the system safety process on the product line programs, and this is less favorable for developing a safety culture in comparison to the other alternatives. Program personnel may perceive the tasks supported by the safety experts to have a lower priority than the other program tasks. Also, system safety may not be perceived as the functional manager's highest-priority responsibility. The functional manager may have been assigned responsibility for the safety experts in addition to other domain responsibilities so that product line leadership can avoid another direct report. The manager is assumed to have earned his reporting status due to expertise in a domain other than system safety, since there is no career path leading to a management position for system safety in this organizational structure. System safety prioritization affects the development of a safety culture. Direct executive action to prioritize safety tasks can mitigate adverse effects of this organization on the safety culture.

6

Safety Lifecycle

Background

Chapter 3 discussed the automotive safety lifecycle as it applies to a series production automotive product. This is supported by the lifecycle envisioned by ISO 26262 and is specific to the automotive safety lifecycle rather than the more-generalized lifecycle in IEC 61508. It is generally accepted across the automotive industry that system safety must be designed into the product before production launch rather than added to the product after it is built. System safety is considered in the earliest concept, as expected from applying the safety culture discussed in Chapter 5. Even before a contract is granted, aspects of system safety are included in the request for quotation and the subsequent quote. Then, system safety must be ensured throughout the design and development and maintained in operations (production and repair), use, and disposal. The system safety process is used to ensure that each of these considerations is adequately addressed.

The phases of the automotive lifecycle consist of concept through disposal. Auditing and assessment ensure that the system safety process is executed, using a systematic approach in each phase. This is mandated by the safety policy and implemented by the safety process. Any omission could lead to unreasonable risk, so diligence is required.

Consumer expectations for safety demand this diligence. ISO 26262 provides some guidance regarding systematic execution of a safety process by phase. Work products are provided as artifacts that are assessed to ensure that a systematic approach is used for functional safety. Chapter 3 discusses how this approach can be broadened to include SOTIF. As discussed, the content of the work products is expanded to include SOTIF requirements consideration.

Each lifecycle phase has different activities and tasks to be executed. The system safety process can address all of these phases so that execution of the activities and tasks is monitored and ensured. They are included in an overall development and launch process that includes identifying and providing for maintenance and disposal. In this way, system safety for post-launch activities can be provisioned and reviewed prior to product launch.

Each phase may provide prerequisites for the following phases. Such prerequisites can be anticipated by the system safety process. Overall assurance of process execution with support from system safety experts in an enterprise with a positive safety culture can demonstrate diligence in ensuring system safety. Senior enterprise leadership, the system safety

Automotive System Safety: Critical Considerations for Engineering and Effective Management,
First Edition. Joseph D. Miller.

process, auditing, assessment, and system safety metrics and review all target this objective. The actors in each phase participate in achieving system safety.

Concept Phase Safety

Preliminary Hazard Analysis

The concept phase of an automotive safety product may start before a VM issues an RfQ to potential suppliers and before an initial response is provided. The VM may have considered recommendations acquired from the commercial domain, such as marketing, concerning consumer demand for feature content in vehicles. This may then have been developed into a preliminary set of requirements sufficient for a preliminary hazard and risk analysis (PHA). It is important to have an early system description – or item definition, per ISO 26262 – to perform a PHA. The boundaries of the system, its function, and an estimate of the authority to perform these functions need to be considered. Concerning authority, for example, consider the amount of steering handwheel torque required to overcome unwanted lane-centering assistance. If the torque required is 2–3 newton-meters (NM), this is significantly less authority than 6–10 NM. The potential hazards may be different.

The PHA allows the VM to include an early estimate of the properties of the interfaces and integration so that related requirements for supplier support are included in the RfQ and subsequent quote evaluation. These requirements may include monitoring that an interface signal is within limits, and requirements concerning what action to take based on this monitoring. Because the PHA also provides automotive safety integrity levels (ASILs), the ASIL attribute of these safety requirements – where D is the most stringent and A the least – may also be included in the RfQ. Supplier selection may include consideration of whether potential suppliers have the capability to meet these ASILs. The VM may have previously prepared such requirements based on the safety classification of the system as a separate document and can include them by referring to this document along with other unique system safety requirements. Another document to be referenced may list the expectations for a joint development similar to the development interface agreement required by ISO 26262. If the product falls within the scope of ISO PAS 21448, then the interface expectations for this joint development, such as validation expectations, can also be included. The supplier is expected to review and respond to these expectations in the subsequent quotation. This can support consistency of the RfQ, and omissions are less likely.

The supplier can also perform a PHA and use this to tailor the activities included in the supplier's system safety process in order to support a cost estimate for the quote. If no ASILs have been provided to the supplier by the customer in the RfQ, then the supplier may determine these independently in order to anticipate the scope of the safety process to be employed. The PHA may also be used to review the requirements provided by the customer. The supplier may anticipate additional scope of requirements for additional interfaces, ASILs for additional hazards, or differences in ASILs or undesired behaviors of intended functions. For example, consider a column lock system where unwanted locking of the column while the vehicle is moving is not identified as a hazard. The supplier has performed a PHA and identified this as a hazard with ASIL C (based on data) or D (to meet customer requirements).

The supplier intends to meet this ASIL if awarded the business, since the supplier has supplied the system meeting ASIL D requirements before. If such discrepancies or conflicts arise, clarification can be sought so that the quote is based on a common understanding. The supplier may choose to quote based on the higher ASIL in order to produce a product that does not increase the risk to the general public. This may be an ethical decision that is supported by the supplier's enterprise leadership. It is expected that the VM will agree, unless there is some other mitigation of this hazard in the application that is unknown to the supplier.

This agreement has always been important for functional safety, including agreement regarding the ASILs of various potential hazards. Rigor and associated resource consumption may increase dramatically as the ASIL of requirements increases, and the cost of safety mechanisms and other architectural features may increase as well. Therefore, it is important that the agreed-on ASIL not be too low, to protect the general public; and not having the ASIL too high is important, to avoid reducing the availability of safety-enhancing systems by driving a higher cost. It becomes even more important for SOTIF-related products where the costs of completing the activities of areas 1, 2, and 3 of ISO PAS 21448, discussed in Chapter 3, are being quoted.

It is important to agree on what is needed for validation and how it is achieved. The supplier may have completed some of these validation needs already for the product. The methodology used needs to be transparent and agreed on with the VM. The efficiencies discussed Chapter 3 also should be agreed on: combinations of driven miles, expert judgment in selecting worst-case routes, and simulations may have been used, and such efficiencies are considered in the quote. Completion of the item definition enables this because the hazards can be determined. Then validation of hazard avoidance can be planned.

The PHA can be performed efficiently by using a hazard and operability study (HAZOP) on the external interfaces of the safety-related system. This includes not only the actuator, but also communication channels and the human-machine interface (HMI). The HAZOP guidewords should be functional in nature and relate directly to the actuator and data interfaces: for example, *too much, too little, in error, missing, wrong timing, too early, too late, too often,* and *not often enough.* Consider, for example, an electric steering system: the actuator executing in error includes assist in the wrong direction and assist when not requested. This is included in the HAZOP analysis, and the effects can be evaluated. Both of these errors can cause a random accident that is an ASIL D hazard in an annex of SAE recommended practice J2980 by consensus, not based on traffic statistics. (This annex was compiled by having VMs and tier 1 suppliers list potential hazards that they agreed on, along with the range of potential ASILs they agreed on.) Failure of the actuator to provide steering torque when requested can produce a similar hazardous event in a SOTIF-related application, as does failure to recognize a request for steering torque. Considering the authority of the actuator can lead to identifying this hazard.

Preliminary Architecture

Additional guidewords are used to continue the HAZOP as applied to the functions of the product under consideration. The expectation is that this systematic approach will provide confidence that the hazards elicited in this manner are complete. This approach adds credibility to the resulting PHA – the HAZOP documentation shows diligence.

During the concept phase, a preliminary architecture can be developed. The boundaries of the system are known, as are the intended functions at the interfaces. The architecture can represent an approach to achieve these functions that guides the subsequent design. This architecture is composed of functional blocks and functional requirements for each block. The safety requirements traceable to the safety goals also need to be assigned to these functional blocks. Safety requirements for this architecture can be derived by performing systematic analyses on the architecture, which will increase confidence in the completeness of the architectural requirements elicited.

Some examples of systematic analyses include a fault tree analysis (FTA), an event tree analysis, diagnostic coverage, and analysis of requirements from other systems. Each of these types of analyses has advantages associated with it that differ from the others. The FTA is especially useful because it allows consideration of combinations of failures that can lead to the hazards identified by the PHA. Then, the possibility of common-cause failures leading to these combinations is evaluated using a deductive method such as a failure modes and effects analysis (FMEA). If a single failure can lead to a hazard, an evaluation is made to determine whether the likelihood of such a failure is sufficiently remote, such as a mechanical failure that has enough design margin of capability beyond the requirements of the product. Consider, for example, the rack of a rack-and-pinion steering system. It is tested to three lives: testing that exceeds the requirements of the product over a single lifetime by a factor of 3. This ensures that the probability of a serious failure of the rack to cause disconnected steering is not credible. If the single failure is not sufficiently remote, then the architecture is reconsidered to build in redundancies to mitigate the risk of this failure and ensure safe operation in case of such a failure. This is a fail operational state, and it may include a minimum risk condition. The FTA is updated to show these redundancies.

Part 3 of ISO 26262 notes that the FTA and HAZOP can be used to elicit system requirements. While the HAZOP technique is well known for use in analyzing operators in manufacturing plants or other facilities, it also is applicable for functional analysis. The HAZOP technique is referred to as an event tree analysis (ETA) when applied qualitatively to failures in the architecture, although the term *ETA* also refers to a different type of analysis.

The ETA is an effective method for determining architectural requirements systematically. Determining diagnostic coverage at the architectural level is not suggested as a technique for eliciting system requirements, although this was discussed when the first edition of ISO 26262 was being developed: it was considered clever and a good way to elicit requirements for software-based safety mechanisms early in the program. This is especially useful for software because software development is often the most labor-intensive activity on the program. The advantage of the ETA for the safety process, using only estimates of failure rates at the architectural level, is to ensure that the required single-point failure metric (SPFM) is achieved at the hardware level when the detailed design is available and the SPFM is computed. Refinements to the safety mechanisms may be needed, but this should be a rare occurrence. If the architecture has been used previously and the hardware design achieved the requisite SPFM, then a diagnostic coverage analysis at the system architectural level will not be beneficial. The previous diagnostics can be assumed to be sufficient as long as there are no changes affecting the hardware design. However, for a new design, it may improve the likelihood of a successful design. The coverage at the architectural level is likely to be a good estimate of the diagnostic coverage at the hardware level. Similar benefits are achieved by a qualitative system-level FTA, provided this FTA includes the envisioned diagnostic with sufficient diagnostic coverage capability. Then, the specified diagnostic requirements are included in the architecture and in the software as required. The ETA can also be useful in this regard, provided diagnostic requirements are included. Both elicit requirements for diagnostics that are captured within the architectural requirements, and hardware and software requirements follow.

Requirements

When these systematic analyses are completed, there can be confidence that the system safety requirements are complete. These requirements include the safety goals determined from the PHA, and the top-level safety requirements at the system architectural level that include the top-level hardware requirements and the top-level software requirements. To achieve such confidence requires review of the analyses, such as an independent peer review that includes experts from each of the domains responsible for complying with the architectural requirements as well system engineering experts. The review can also be performed using simulation of the requirements and the response to inserted faults to determine whether the faults are properly managed. At the architectural level, the simulation is realized using behavior models of the architectural blocks that simulate compliance with the architectural requirements. Review of the architecture using the simulation results is more efficient.

Any revisions to the safety requirements are then assigned to architectural blocks, and the domain experts responsible for these blocks can agree to comply with these assigned requirements. These architectural blocks also include system functional requirements. System safety requirements are not in conflict with these functional requirements and are easily recognized. They should be uniquely identified, atomic, and verifiable according to ISO 26262, which also requires an ASIL and a status. SOTIF safety requirements have similar attributes but may not necessarily have an ASIL, since ISO PAS 21448 does not require ASIL assignment.

The safety requirements and architecture are hierarchical, to avoid complexity by controlling the number of blocks at each level of the hierarchy. Sometimes the number 7, plus or minus 2 [8], is used to determine the best number of blocks at each architectural level for understandability. This has to do with experiments concerning human one-dimensional absolute judgment and memory span, which are adversely affected by longer lists and a greater number of different stimuli. In this requirements hierarchy, lower-level safety requirements satisfy the parent safety requirements.

For example, consider a top-level hardware requirement that no single-point hardware fault may lead to a violation of a safety goal. This is a parent requirement. The individual safety mechanism requirements to ensure this, elicited by the SPFM, are child requirements to satisfy this parent requirement. For the design, these safety requirements are divided between hardware and software safety requirements. For instance, some safety requirements elicited by the SPFM and latent fault metric (LFM) are hardware requirements or software requirements. In addition, safety requirements are assigned to other systems to document assumptions concerning the behavior of these systems to achieve safety. A radar cruise control system may depend on a requirement to the engine management system to recognize whether a cruise control request is out of bounds or flagged as invalid. This supports subsequent requirements verification and traceability.

Design Phase Safety

When the system requirements have been elicited, represented in an architecture, and reviewed, they are used to support the design of the system, software, and hardware. This mitigates the risk that is envisioned if, due to schedule urgency, the software or hardware design must start based on what the system requirements are expected to be by the software or hardware design group. Starting without the prerequisite requirements available is similar to designing a safety element out of context. The risk is that part of the design may need to be modified when the system-level requirements become available, or that content will be included that is not based on the prerequisite requirements and is not required. This unrequired content may cause an unexpected behavior anomaly that is detected late in, or after, the program product validation.

It is advantageous for the physical system design to be made consistent with the architecture. For example, this avoids the potential inefficiency of a purely functional architecture that is not related to the physical design. While a purely functional approach may be useful analytically, the functions must be sorted by the domain responsible groups so that analysis of the physical design is executed without the benefit of reusing and expanding the top-level analysis that has been performed on the architecture, because it may not apply directly to the physical design. This consistency allows direct applicability of the architectural analyses to the design.

Design-Level Safety Requirements

Top-level requirements are derived for the system's physical partitions, such as the electronics control unit hardware, sensors, and software. The design can then satisfy the system safety requirements directly. These become the parent requirements for each partition of the design. In addition, detailed hardware design safety requirements can be derived that are traceable to the system safety requirements assigned to hardware. These flow down to elements of the hardware, such as the voltage regulators.

As the hardware design is developed, further analyses are performed to elicit additional safety requirements. For example, the hardware-detailed SPFM and latent failure analysis are performed to document that the design implemented is compliant with ISO 26262 requirements; they may also elicit additional requirements for software for safety mechanisms required for this ISO 26262 compliance. In addition, common-mode and common-cause analyses are performed, taking into account the actual selection of components. For example, the use of identical components from the same physical lot of parts from a component supplier can pose a vulnerability to common-mode failures. The design needs to consider the effect of failures due to fluid ingress – into vulnerable mechanical parts or electronics, which may cause shorts in unprotected components – so that, for example, the first components that can short are predictable and triggers a safe state or a minimum risk condition. Potential toxicity hazards due to leakage also need to be evaluated if applicable; measures can be taken to contain or eliminate toxic material and reduce this risk to an acceptable level.

The hardware design can be updated to comply with any additional system safety requirements elicited. For example, additional redundancies or other measures against component failures may be required, such as hardening against potential cascading failures. If necessary, the software requirements or system-level system safety requirements are modified: for example, following a change in the scope of the system functional requirements. All modifications are documented to support traceability.

Traceable, detailed design software requirements are derived from the system architectural requirements assigned to software in a manner similar to the way detailed design hardware requirements are derived from the system safety requirements assigned to hardware. This can be performed more efficiently if the system architectural blocks for software match the partitioning of the software design, because the top-level software detailed design architecture will be compatible with the system architecture. Otherwise, interpretation and adjustments will be necessary, to link the system-level requirements to the top-level software requirements to achieve traceability.

Then, the top-level software requirements are assigned directly to the detailed software design architecture. This architecture can be made hierarchical in order to reduce complexity while improving comprehensibility. Further safety requirements that emerge from the software design are elicited by performing system-level software safety analyses on the design. This analysis should be systematic, such as a software FTA if focused on a particular hazard, or other specialized analysis such as an ETA, which is discussed next.

The software architecture is analyzed with an ETA in the same way discussed in the concept phase for analyzing the system architecture. The interface of every software architectural block is evaluated using guidewords, such as *missing, in error, too early, too late, too often,* and *not often enough.* If there are safeguards in the hardware, then some guidewords are removed, such as *in error* if the processor is a protected lockstep that detects computational faults so that a safe state is achieved. The software design architecture ETA elicits requirements systematically to improve the confidence that the software safety requirements are complete. This is useful to elicit safety requirements to cope with the software reaction to the exception of a hardware failure, as well as requirements to cope with deadlocks and livelocks, regardless of cause. If software is reused, then only the changes may need to be submitted for this analysis.

The software design is updated to comply with any additional system safety requirements elicited. For example, if failure to execute is identified as an unsafe failure for a particular software routine, then this routine is made subject to execution monitoring, perhaps using a feature that exists in the safe operating system, if used. If necessary, the hardware requirements or system-level requirements are modified. For example, a requirement may be elicited for hardware memory partitioning protection that is made available by the selection of the microprocessor version. All modifications are documented to support traceability; this traceability is enabled by a configuration control process. A mature software process is used to control systematic software errors. To facilitate this control, engineering quality may audit the software process and support with metrics the achievement of higher maturity levels. References for these maturity levels can help, and can be found in standards derived from ISO/IEC 12207. Diverse software may also be used, but that is not discussed here.

After the design requirements are derived and the design is complete, it is verified to meet the system safety requirements. A verification plan is made that ensures each and every safety requirement elicited will be verified. Appropriate, effective methods are chosen for verification and linked to each requirement being verified, for traceability. Methods are chosen based on ASIL for the functional safety requirements, if ISO 26262 is being used. Alternatives can also be chosen to replace or supplement these methods. SOTIF-related requirements are verified using methods consistent with ISO PAS 21448; these include verifying not only system performance but also sensor performance and limitations.

Verification

Appropriate scenarios and conditions are used for tests, or simulation can be used to verify known use cases prior to validation of unknown cases. Efficiency is gained by using edge and corner cases to bound an equivalence class of scenarios and reduce the number of use cases. Details of the verification, such as the test cases, are identified for each requirement and linked for traceability.

After these verification methods are documented in the verification test plan, along with pass/fail criteria, the verification is executed. This is often accomplished by an independent group within the enterprise, which is supported by external resources, if necessary, to allow for the required scope of verification. System verification is executed in a hierarchical manner, from bottom to top; this helps ensure that it includes subsystems that are functioning as intended, to avoid unnecessary repetition of invalid higher-level tests.

The result of each executed verification method is reviewed to determine whether the pass criteria have been met, and the result of this review is recorded. Care must be taken to ensure that the verification method is valid and its execution is correct, to avoid misleading results. Corrective actions are implemented and documented to resolve any nonconformance. A systematic review of the status of all requirements is performed to ensure completeness of verification. Metrics may, and perhaps should, be kept and reviewed on a periodic basis to determine the progress of verification, identify any issues, and ensure completeness of verification of all safety requirements. Some requirements databases support the use of automation to help determine status; this simplifies the review process and supports systematic review.

Manufacturing Phase Safety

The intent for system safety in the manufacturing phase is to retain the safety designed into the automotive safety-related system or component throughout production and assembly and into the product when it is manufactured. The safety of the design is verified and validated prior to release for production: the safety case is assessed to be sure there is sufficient evidence to support series production and public sale, including evidence compiled to ensure that design safety is retained in production.

Again, systematic analyses are performed to elicit the complete set of system safety requirements. This includes analyses to determine specific production considerations for the design and to evaluate the suitability of production processes for the design. The design failure modes and effects analysis (dFMEA) can be used to identify and evaluate critical characteristics that need to be controlled in manufacturing: they are noted on the production drawings and specifications before those are released for production. Then the manufacturing process is created, perhaps by updating an existing manufacturing process, taking these critical characteristics into account with appropriate process controls. The drawings may need to be released before design verification (DV) is complete, but after the critical characteristics are identified, in order to enable selection of capital equipment and tooling.

The process FMEA (pFMEA) can also be used to elicit safety requirements that mitigate errors that may affect safety and could occur in executing the process. These mitigations are included in the control plan derived from the pFMEA. Design issues that affect the process may also be given as feedback to the product design organization if these requirements cannot be achieved. New redesign requirements are then defined to mitigate these issues, enabling the product design organization to implement appropriate changes in the design and re-verify any other existing requirements that are impacted by the change.

The design changes and manufacturing safety requirements are documented and made traceable to the manufacturing process, including configurable software that is specifically discussed in ISO 26262. Verification of compliance with these is linked to each of these requirements. A production change control system should be established that is assessed and audited to ensure that the safety requirements are always considered whenever a manufacturing process change is proposed. This process is combined with the change control process used in development for economy of software tooling for the enterprise. This way, it can be ensured that the changed process is as safe as the process it replaces. This is necessary to avoid unreasonable risk to the general public: no unreasonable risk is introduced, and system safety is retained.

Safety in Use

If safety of the design has been assured throughout development, and this safety has been maintained in manufacturing, it is expected that the product will be safe in use. This safety in use includes not only the vehicle operator but also all other road users exposed to the product when it is in use. This is true of all safety-related automotive products, regardless of whether ISO PAS 21448 is applicable, including the safety of the human-machine interface as well the safety of the system function and the safety of the system in failure.

For example, consider an electric steering system that is replacing a hydraulic system. All these considerations come into focus. The human-machine interface is basically the same except for the way the system responds: when a hydraulic system starts or stops providing assist, the hydraulic fluid dampens the sudden appearance and loss of assist torque on the handwheel. This somewhat softens the interface from the system to the human and does not startle the driver. An electric steering system does not have this hydraulic damping, so the feel could be somewhat harsher than expected. Thus, the electric system needs to be evaluated under different use conditions to determine whether the application and removal of assist torque presents a potential hazard. When the system starts, the assist may become active more quickly than the operator can react, and the handwheel could move without the driver intending to move it. One use case is when the system is energized at a standstill and there is residual torque that may cause handwheel movement. Could this injure the operator? It is a rapid movement, but easily controllable. Also, consider when assist is removed due to a failure or reinstated while driving: some softening of the start and perhaps at stop may be added, to avoid startling the operator.

Likewise, for SOTIF-related systems, the requirements of ISO PAS 21448 are considered for systems in scope and systems with higher levels of automation. Operation in these cases should also ensure that the driver has a correct understanding of the system's capabilities as well as validate that these capabilities are sufficient in all the intended use cases. These considerations are discussed in Chapter 3 and include safety in potentially ambiguous scenarios as well as foreseeable misuse. Cases of abuse may be considered out of scope but can be taken into account when it is practical to do so. Simulation or actual testing when possible provides sufficient evidence for simpler systems. Sometimes these simulations are run with naïve participants included in simulated driving to evaluate the interface. Validation may increase with the scope of the automation and may include extensive driving in urban and highway use cases in many countries under varying environmental conditions. A great deal of data may be collected to provide evidence for evaluation.

Safety in Maintenance

Safety in maintenance considers both the user's manual as well as the maintenance manual, because both contain information related to maintaining the vehicle and vehicle safety systems. These are supplemented with messages to the driver and signals that are easily understood. For example, in the user's manual, warnings are included about potential misuse as well as expected servicing requirements. Vehicle signals and information may also warn of these considerations as reinforcement.

Consider the example of a convenience system such as adaptive cruise control that executes longitudinal control correctly under most nominal conditions. The user's manual explains any limitations such as adverse weather, driving conditions, and non-reaction to stationary objects. The user is advised to read the manual by the salesperson at the time the vehicle is acquired. Reading the manual calibrates or sets the user's expectations appropriately for the scope of the system's operational capabilities. However, the user may choose not to read all the information in the owner's manual. If the vehicle has a message center, then this message center may further reinforce the user expectations derived from the manual during or before system use. This message center information supports awareness of the intended use: it is a reminder, but not enforcement. The user will learn from experience.

To support installing a vehicle safety-related system safely after manufacture, the maintenance manual has installation instructions with warnings including safety-related consequences of mis-installation. Pictures, explanations, and colored symbols can be used.

For example, consider a safety-related automotive system where misalignment of the radar can lead to potential hazards: vehicles in the correct lane may be poorly monitored because the path is misinterpreted. Instructions explain how to align the radar, and a warning indicates that misalignment of the radar may cause inappropriate braking, or a lack of braking or acceleration when expected by the user. For every warning, an explanation of the consequences is included.

Also, consider an example where torque applied to bolts in the steering intermediate shaft is critical, and over- or under-torque may lead to disconnected steering. The bolt may release or fatigue over time. The manual includes a warning and specification that explain this criticality, along with instructions for applying the correct torque. Also provided are diagnostic information that explains maintenance-related tests and maintenance messages, and explanations of the repairs and consequences that may otherwise result.

Sometimes, system replacement is necessary for safety, instead of repair, so this is also made clear in the maintenance manual with appropriate warnings. Removal and replacement instructions are included, along with associated warnings and cautions and the consequences these warnings avoid.

Some foreseeable maintenance anomalies may warrant special instructions: for example, what to do if a safety system is accidently dropped. The potential consequences of damage are explained. It may be possible to check for damage, or perhaps it is recommended that the system not be used under certain circumstances even if damage is not apparent. The system may be compromised internally, depending on the severity of the drop. Clarity is important for safety, so clear criteria must be established.

In addition to maintaining safety of the system, safety of maintenance personnel is taken into account during preparation for system maintenance as well as maintenance activities for safe disposal. In particular, preventing injury from contact with stored energy may require measures to ensure that maintenance personnel are not injured by its release. For example, consider a system with a pressure accumulator: instructions may include a safe way to relieve this pressure. Likewise, consider the practice of deploying airbags before vehicle disposal: doing so protects personnel from inadvertent release of energy from the propellants. Appropriate warning labels and maintenance manual warnings are employed. Warnings concerning any connections to the pressure or propellants may also have physical warning labels affixed to the system itself.

Inadvertent motion should also be considered: for example, movement of an electric steering system connected to a battery while on a hoist, which may require preparation for maintenance. Disconnecting the power from the electric steering system may be recommended. Warnings are included in the manual and explain the potential danger.

Field data from performed maintenance is helpful to promote continuous improvement. A pareto analysis is performed using this data to focus subsequent developments so they have maximum positive impact on quality and reliability. Collection of this data is especially useful for new systems during initial introduction: such systems do not have the history of other, more mature systems, so these results are unique. Overcollection is advisable until a history is developed to be used to tailor this collection. The maintenance manual may include instructions about what data to record and return. Tooling may improve the efficiency and execution of this collection. This field data collection supports improvement of product maintenance. The data collection is evidence of diligence.

Safety in Disposal

The final phase of the product's lifecycle is normally disposal. This is the product's end of life, though some parts can be reclaimed for use in repairs. Disposal is considered an extension of maintenance, so much of the previous discussion regarding maintenance is also appropriate for disposal. Safety of the personnel involved as well as those who encounter the product after disposal is the primary concern.

Instructions for disposal are provided in the maintenance manual, including warnings that clearly explain the potential consequences if the warning is not heeded so that the priority of following these instructions is understood. For example, it is appropriate to discharge stored energy prior to disposal in order to prevent potential injury while reclaiming other nearby components or systems. For example, when disposing of a vehicle, the airbags are discharged; otherwise, inadvertent deployment may harm personnel reclaiming components near the airbag system.

Some systems can be salvaged as used equipment, so any safety requirements should be considered, such as requirements to prevent an inappropriate installation due to calibration for another vehicle. For example, if a stability control system (SCS) is calibrated for a particular model vehicle with specific content, installing it in a different model may cause the occurrence of inappropriate interventions by the SCS, applying brakes to individual wheels and causing the vehicle to change direction. In these cases, design measures may possibly be included to support proper calibration. The useful lifetime may also be specified. This warning not to exceed the useful lifetime should be included.

7

Determining Risk in Automotive Applications

Since *safety* is the absence of unreasonable risk, it is logical that the risk must be determined in order to determine whether it is unreasonable. This may not necessarily be a numerical value, but it should be actionable in order to mitigate unacceptable risks and make them acceptable.

In ISO 26262, *risk* is considered the combination of the probability of occurrence of harm and the severity of that harm, where *harm* is damage to persons. Probability of occurrence and severity are not used to calculate a risk value in ISO 26262, but are estimated separately and combined with controllability to determine an automotive safety integrity level (ASIL). This ASIL is used to index requirements from the standard. In order to determine the occurrence of harm, it is necessary to know how, and under what circumstances, that harm can occur. For example, if a system can cause a random accident on a highway, and the harm is death, then statistics can be found for this probability, or it may be estimated. For an automotive system, this process leads to an evaluation of what the system can do when it fails – and when it is operated correctly, even if misused – to cause harm. This extreme operation may not be the system's expected functionality but could be if the system contains some automated functions.

The risk determined as a result of failure due to a fault in the system is evaluated using ISO 26262 and assigned an ASIL. Then, the probability and severity of harm are evaluated. The risk determined as a result of the system operating as intended, including misuse, is evaluated using the methods in ISO PAS 21448; this may require more extensive evaluations to be used. Both standards may consider traffic statistics. Then risk reasonableness can be determined.

Analyze What the Actuator Can Do

Imagine an automotive system that is intended to do nothing and can do nothing, even in failure. Such a system is completely fictitious and does not exist in any practical application. It would be completely useless: it would have no ASIL and would not increase the risk to the general public. The system cannot cause any harm while in use because it cannot do anything to cause harm. When it fails, the failure does not remove any function or cause any perceivable effect. Such a system has no authority, and there is no unreasonable risk.

Automotive System Safety: Critical Considerations for Engineering and Effective Management,
First Edition. Joseph D. Miller.

As the authority of an automotive system increases, so does its potential for risk. This authority gives the system an effective function that may have the capability to cause harm. This harm could occur in normal use or as a result of a malfunction. Consider a lane-keeping assist system that can provide up to 3 Newton-meters (NM) of steering torque to assist in keeping the vehicle in the lane. This is the torque as measured at the handwheel that is applied automatically to correct the position of the vehicle in the lane while the vehicle is moving. In this example, tests with a broad sample of naive subjects were conducted using the intended vehicle and confirmed that the driver could always keep the vehicle in the lane even if the 3 NM of torque was applied in error and without warning. No driver was startled into losing control, and the tests were run in accordance with accepted practices [9]. Such a system is assigned QM (quality management, no ASIL per ISO 26262) for the hazard of wrong assist if controllability is judged as C0 (controllable in general per ISO 26262). This is the case if experts consider the possibility of someone not being able to control the vehicle as negligible. If controllability is judged as C1 (simply controllable per ISO 26262), severity S3 (life-threatening injuries [survival uncertain] or fatalities per ISO 26262), and exposure E3 (medium probability per ISO 26262) for the condition where the car slowly drifts into the lane of oncoming traffic and a head-on collision cannot be avoided, an ASIL A could result. Experts may consider this case if it is deemed plausible or may not if it is considered that the driver should see the oncoming traffic and is perfectly capable of avoiding it. The system is intended to be used with the driver's hands on the wheel and the driver's eyes on the road. Such a message is displayed before driving, and this message may be reinforced with audible signals. Misuse would be if the driver's hands were not on the wheel for an extended period. While the system is able to maintain lane position, higher torque curve requirements and other emergency maneuvers may require driver support. The system is designed to detect this hands-off condition by means of handwheel torque sensing, perhaps reinforced with direct touch sensing built into the handwheel. These methods are designed to be generally effective against foreseeable misuse and are difficult to abuse, particularly if both are used redundantly. Warnings and shutoff then mitigate this misuse. After a short time of detection, the warning is given; and then, after a short time of warning, the lane-keeping function is disabled so the driver must take over. Again, tests with a broad sample of naive subjects confirm that the warning and shutoff strategy is controllable in accordance with the Response 3 Code of Practice for an advanced driver assistance system (ADAS) [9]. Such testing in accordance with this global guidance provides evidence of diligence and satisfies area 2 of ISO PAS 21448. Satisfying area 2 is enough, and the risk is not unreasonable.

Now consider the same system, but with full steering authority available, and without the expectation that the driver's hands are always on the handwheel. This is sufficient authority to steer the vehicle during all driving situations encountered in normal use of the vehicle on public roads. Additional torque input from the driver is not required in any foreseeable use case. This authority is judged the same as the steering system previously discussed in Chapter 6, regarding the concept phase: the hazard analysis for a manually operated power assist steering system is directly applicable to this automated steering system. The authority of the system is the same over the directional control of the vehicle. An ASIL D is assigned to the potential hazard of unwanted steering torque caused by a fault of the system. This is the consensus expressed by the vehicle manufacturers during the discussions of the annex by the Society of Automotive Engineers (SAE) J2980 [6]. The safe state or minimum risk condition requires a redundant means of steering, considering the requirements for a safe transition. The capability for a fail operational state is included in the design of the steering system; this provides sufficient steering capability to maneuver the vehicle while it is moving, in order to achieve a minimum risk condition. The intended function of the system is to follow the lane under all conditions specified. Examples of specified conditions include highway driving on freeways, city driving, and campus driving under restricted operating conditions, such as fair weather. The system has safeguards to prevent use in unspecified conditions by confirming sensor confidence, perhaps supplemented by geofencing. This can prevent the initialization of a journey when the system is not completely functional due to internal issues or external issues. Now, when safety of the intended function (SOTIF) is evaluated using ISO PAS 21448, all three areas are appropriate. Analysis and testing of known use cases in areas 1 and 2 are needed, as is validation, including searching for unknown use cases. Efficiencies are sought using both simulation and equivalence classes. The use of geofencing may also reduce the scope of use cases needed for validation. A minimum risk strategy must be determined and evaluated, because the driver cannot take control if the system does not engage. The system may consider speed reduction, perhaps including a controlled complete stop, or steering to a safe space, or completing the journey, depending on the reduced capability caused by the failure. (Full validation is discussed in Chapter 3.) Full validation achieves no unreasonable risk.

Analyze Communication Sent and Received

In addition to the risk evaluated by the system's actuator, the risk caused by other systems due to miscommunication must also be considered, along with the risk of mishandling communication from other systems. Such communications are often critical in safety-related systems such as steering, stability control, and engine management, because these messages can influence the motion of the vehicle: the information is included in the execution of the algorithms used to determine the control of the system's actuators.

Consider again the system that can provide up to 3 NM of steering torque to assist in keeping the vehicle in the lane. The steering does not sense on its own the position of the vehicle in the lane and the torque needed to correct it; input comes from another system supporting the lane-keeping system. The lane-keeping system is composed of a camera that communicates with an electric steering system. The camera signal to the electric steering system provides data about the requested magnitude and direction of the steering torque. Even if the communication to the electric steering system is in error, the error cannot exceed the 3 NM authority, due to a limitation imposed by the electric steering system. The electric steering system either limits the torque to 3 NM or cancels the torque command and alerts the operator of a fault, depending on the system's design and strategy. The risk evaluation of the camera system does not change due to communication; the camera has no more authority than 3 NM, so the risk due to a camera communication error is no greater. The burden is on the electric steering system to limit the authority to no more than 3 NM. However, if the 3 NM limit were imposed by the camera, then the risk of this communication error could be ASIL D, just as the unlimited case. Even if the camera computes the limit correctly, a communication error could cause an increased value to be commanded. Likewise, the communication channel of the electric steering is separate from the 3 NM limiting function. The possibility of a common-cause error is mitigated by this separation. This separation imposes no new miscommunication risk. The requirement on the steering system to limit the camera authority has an ASIL D attribute.

Now consider the same system, but with full steering authority available, and without the expectation that the driver's hands are always on the handwheel. The highly controllable 3 NM limit is no longer in place, so the hazards of the steering system determine the hazards of the lane-keeping system. The risk of a miscommunication by the camera is no longer mitigated by a torque limiter in the steering system, so an error in communication can command full unwanted torque in the steering system. Corruption of the data can be mitigated by a properly designed checksum and alive counter. The checksum provides redundancy on the data by appending redundant data – a decodable extension – to the message. And the alive counter mitigates stale data by including a rotating message code: it updates on each message to mitigate a "stuck." Detection may prevent such an invalid command from being executed. However, this is predicated on the availability of a suitable safe state or minimum risk condition. Not executing the incorrect command and switching off the system is not a safe state for this system. Such a minimum risk condition may require an additional communication channel, perhaps from the same or a different sensor; then arbitration may be necessary to switch to using the other communication channel. Likewise, errors on the steering end of the communications channel are not mitigated by a torque limiter. These errors must be taken into account, such as by periodic testing, error detection, or redundancy, and a safe state or minimum risk condition achieved. This may require additional redundancy in the steering system itself. Higher-level system safety requirements are like the lower authority system, but lower-level implementation requirements are updated. SOTIF must be validated for this higher authority implementation.

Determine Potential for Harm in Different Situations and Quantify

In general, automotive safety-related systems do not have a direct potential to cause harm themselves. Inadvertent actuation of a steering system does not directly cause harm to the occupants of the vehicle or to other road users. The harm is caused by the behavior of the vehicle in situations where harm could result. If the vehicle is unavoidably steered into an object at high speed, then harm could occur to the vehicle occupants.

An exception is the inadvertent release of energy, such as by an airbag passive safety system. An airbag system may make direct contact with the vehicle occupants and has the potential to malfunction and cause harm. If too much energy is released, harm may result in the form of too much force from the bag, or too much force causing failure of components and injuring the vehicle occupants with the resulting shrapnel. In general, however, the greatest concern for an airbag system is inadvertent deployment of the driver airbag, due to concerns about loss of vehicle control; and failure to deploy without warning, due to potential injury in the relatively unlikely event of a collision with enough force to warrant airbag deployment. The deployment of the driver-side airbag inadvertently will at least startle the drive, and it may cause the driver to release the steering wheel. If the driver is reaching across with one arm to steer, the airbag deployment may drive that arm to the driver's head, causing a fractured skull and breaking the arm. This may lead to a general loss of control of the vehicle. Thus, to determine the potential for harm, evaluation of the different operational situations that the automotive safety system may encounter is required.

Traffic statistics indicate that the probability of serious injury or death is less than 10% for a general loss of control. Nevertheless, many vehicle manufacturers (VMs) consider this S3 harm for determining an ASIL. In ISO 26262, being in an operational situation with the potential for harm due to a failure is referred to as *exposure*. In the case of inadvertent airbag deployment, all operational situations are considered. This leads to an exposure of E4, which is evaluated to determine ASILs. SOTIF also considers these situations.

Exposure

Limiting exposure to dangerous situations is one way to reduce the risk potential of an automotive system. In the example of inadvertent airbag deployment, there is exposure to all potentially dangerous situations. As discussed previously, if a system has no potential to do anything and is not intended to do anything, then it has no risk. This is an extreme and fictitious example of reducing risk, but practically, some limitation is useful. By limiting the authority of the actuator for some features, the potential risk of those features is also reduced.

Consider limiting the braking authority or speed reduction of an automatic emergency braking system that is capable of 1 G braking. For example, the braking authority is limited to 0.5 G when the vehicle is above a particular speed. This reduces the exposure to full authority at high speed and associated high-severity rear-end collisions. This braking authority is implemented using the stability control system: it mitigates locking the rear wheels, and the vehicle stays stable.

To determine all the potential situations that are a source of harm for an automotive system, the methodology discussed in [6] may be useful. This is similar to a hazard and operability study (HAZOP) approach with the purpose of determining which situations are potential sources of harm. Here, guidewords are taken from different categories, such as *location*, *road conditions*, and *vehicle state* (each category has a list of guidewords). Then the selected guidewords are taken in different combinations to determine whether there is a potential for harm due to a failure. Inadvertent emergency braking with reduced authority on an icy surface may have a potential for harm, even with reduced authority, if not implemented through the stability control system. As multiple conditions are considered, the potential for correlation is taken into account: for example, ice and low friction are likely to always occur together, so no reduced probability is justified. If conditions are uncorrelated, then the likelihood is less; this may be considered justification for reducing the exposure class for the conditions in combination. Exposure is considered for a particular harm, and the exposure is grouped by severity. Picking the worst-case exposure and the worst-case harm when they are not connected may inadvertently cause an unjustified ASIL. For guidance, consider an appendix in ISO 26262 part 3, where driving on a highway has an E4 exposure but driving on a wet highway is rated E3. This is an example of combining conditions. Further specific maneuvers may additionally reduce exposure.

Exposure is also ranked based on frequency of exposure to situations. This is an alternative to using the percentage of time for exposure determination, such as 10% or more for E4, etc. Frequency enables improved transparency of the risk of exposure. It is used when the operational situation may trigger malfunctioning behavior after a latent fault, rather than just the time exposed to the operational situation. For example, again referencing an appendix in ISO 26262 part 3 for guidance, driving in reverse has an exposure of E2 based on time spent driving in reverse. Exposure to driving in reverse based on time is very low. However, if driving in reverse may trigger malfunctioning behavior for a system, it is ranked E4, because driving in reverse may occur on more than 10% of the journeys considered. For example, if a vehicle must be pulled into a garage or parking place on almost every journey, then exposure frequency is very high. When triggering latent faults is considered, the exposure increases due to the potential time the latency exists. This increases the exposure ranking, and the resulting ASIL also increases.

Priority

This evaluation of situations where harm may occur, which is performed to determine exposure rankings for the ASIL used in functional safety, is also useful when determining scenarios used in validation for SOTIF-related systems. Exposure to potential hazards due to malfunctioning behavior caused by failures within the item is likely to take place in situations where the hazards are due to unintended behavior of the intended function. For example, unintended emergency braking on a dry, straight highway due to malfunction of the automatic emergency braking system may be caused by a failure within the emergency braking system or a limitation of a sensor in the emergency braking system that has not been sufficiently mitigated by the algorithm.

The exposure evaluation is performed systematically, and thus improves confidence regarding completeness. The systematic application of the HAZOP guidewords (e.g. *too much*, *too little*, *in error*) to driving conditions elicits exposure with greater thoroughness than an unguided investigation may achieve. Deeper investigation using combinations further adds to the exposure determination argument for completeness. These situation analysis techniques can have valuable, though limited, usefulness for functional safety but are a rich source of triggering events for the SOTIF evaluation. For example, driving in a crowded parking lot may be low exposure for functional safety, but may be rich in potential triggering events.

Situations that have exposure to higher levels of harm due to evaluation for functional safety are also evaluated for SOTIF, producing supportive evidence of diligence, in contrast to ignoring this information when performing the SOTIF evaluation. These situations are likely to have high exposure for the same hazards caused by unintended behavior of the intended function but may require a focused effort to elicit potential triggering events. The evaluation is expanded for SOTIF by using additional guidewords for the specific function, if appropriate. For example, *misidentifies an object as free space* may elicit a limitation of object-detection sensors that results in classifying the side of a large white truck as free space. Still, efficiency is gained by prioritizing the situations for SOTIF that have the highest functional safety risk. A lack of evidence that this available information has been diligently considered for SOTIF can have an adverse effect on the system safety case. This prioritization of high-risk situations from the functional safety analyses helps guide the situations explored in areas 2 and 3 of ISO PAS 21448. Validation in these areas is labor intensive, and such efficiency may help.

For example, consider the automatic emergency braking (AEB) system discussed in an annex of ISO PAS 21448. This AEB system has authority to apply an emergency braking profile with full braking force. The conditions under which this authority is exercised are limited by an upper boundary of availability based on vehicle speed. Driving routes used for area 3 validation are prioritized and weighted based on the average driving scenarios in the markets where the product is sold. The underlying assumption is that the driving routes used for vehicles having the potential new AEB feature are the same as the driving routes used by existing vehicles that are not equipped with AEB. This assumption supports use of existing data. Scenarios known to have the highest risk for functional safety are to be explored for related unknown scenarios: these unknown scenarios may be safe, or they may be unsafe and require improved handling of the discovered triggering events. For instance, suppose the full-authority AEB system is operating on a straight road during daytime and is triggered in error in a tunnel by an oncoming vehicle that causes an unanticipated reflection. In using higher-priority scenarios, an unknown unsafe scenario is discovered. The specification is updated, and design changes are made to compensate for this unsafe scenario. Credit is taken for this weighting of higher-risk scenarios in order to gain validation efficiency.

Such validation testing is more valuable than repeated testing of known safe scenarios. A consensus based on the significance of these higher-risk scenarios in hunting for unsafe scenarios, and for more heavily weighting the validation to hunt for these scenarios where they are more likely to be found, allows a higher number of validation miles to be claimed based on a value judgment. Prioritizing these scenarios helps reduce public risk and supports safety.

Consider Fire, Smoke, and Toxicity

In addition to the risks discussed in [3] and [4], it is also important to consider the risk of harm that may result from fire, smoke, and toxicity. ISO 26262 discusses only risks from hazards based on malfunctioning that result from a system failure. For example, it does not include fire and smoke unless the function of the system is to prevent fires. ISO PAS 21448 discusses only risks due to the intended function and also does not address fire and smoke. Risks due to toxicity are also not considered potential hazards in either ISO 26262 or ISO PAS 21448.

When choosing materials for products, toxicity is often considered, including materials that are toxic if released into the system surroundings and must be contained. If these materials are acceptable for use, then the design is required to contain them safely and reliably. This containment should be for the life of the product and potentially thereafter. The materials are entered into the International Material Data System (IMDS) and often submitted to the VM as part of the release process.

No restricted materials should be included in the product, including, for example, printed circuit board materials, solder materials, and the composition of alloys used in mechanical parts, such as steel. This enables VMs to meet the required goals of the End of Life Vehicles

Directive of the European Union. The directive has several targets, all dealing with various aspects of the life of a vehicle as well as processing and treatment. It supports safety in disposal: if materials are potentially hazardous to people or the environment when the vehicle is disposed, then the targets are not met. Therefore, the design must be revised to meet the targets.

If there is a possibility of causing exposure to hazardous materials during use or maintenance, requirements to mitigate this are elicited, and compliance is verified. Measures must be taken to either eliminate these materials or contain them safely so that they do not present a hazard to people.

In addition, potential hazards due to fire and smoke are considered both for the case when the vehicle is occupied and also for the case when the vehicle is not occupied. Consider a vehicle that is occupied and being operated. In this case, safety in use is the goal, based on the ability of the operator to control the vehicle. When fire or smoke is in the vehicle passenger area, measures must be taken to prevent or contain the fire and smoke so the fire is not a hazard and the smoke does not reduce controllability of the vehicle. If fire and smoke are outside the passenger compartment, then the fire must be contained to prevent a hazard. Smoke is less of a concern in most cases when the vehicle is moving; it is dispersed by wind and partially contained by the engine compartment or another compartment of the vehicle. The fire is a source of ignition to other systems, so if it is not contained within the system, it is the larger concern.

If the vehicle is not occupied, the potential for harm due to fire and smoke is controlled by externally switching off power to the system. This is executed when the ignition is turned off, perhaps using a relay or other means that may be contained within the system or outside the system. This additional cut-off may reduce this probability to a reasonable level without further consideration within the system, because there is no power to supply the energy to start the fire. Otherwise, additional measures are warranted. Metallic or other nonflammable containment may be used to prevent the vehicle from starting a fire in a structure outside the vehicle and potentially causing catastrophic harm.

8

Risk Reduction for Automotive Applications

History

Risk reduction has been important in the automotive industry since long before the ISO 26262 standard was released on November 15, 2011. For example, mirrors were included to improve the driver's ability to see the vehicle's surroundings. Even the earliest of vehicles had two headlights to allow the driver to see in the dark even if one failed. And bumpers were added to the front of vehicles, along with additional features to improve crashworthiness.

Systematic analysis techniques developed in other industries were carried over to the automotive industry and applied there. The military has used failure mode and effects criticality analysis (FMECA) since 1949 [10], and NASA was using it by the 1960s [11]. Systematic analyses are important for military use in order to reduce the risk of mission failure due to system failures. Such systematic analyses can provide confidence that all potential causes of failure have been analyzed. The intent is to analyze the failure modes of components so that the effects of these failures can be mitigated. These mitigations may include redundancy or a software switch to another mode of operation to complete the mission. Each of these failures can be prioritized based on its criticality: a combination of the severity and occurrence of the effect.

By the 1970s, failure modes and effects analysis (FMEA) techniques were being used in the automotive industry, and they were standardized by the Automotive Industry Action Group (AIAG) by 1993 [12, 13] and then by the Society of Automotive Engineers (SAE) in 1994 [14, 15]. These standards support the extensive use of FMEAs on every automotive safety product. Improvements have been made over the years, and the standards have been updated. Every automotive supplier has its own process to reliably ensure use of the techniques in the standards as they apply to products. Clearly, the need for risk reduction has decades of history in the automotive industry. Some techniques are new, others are not, and advances are still being made.

Analysis of Architecture

While cost is undeniably a major consideration in automotive vehicle design and the award of business to automotive suppliers, safety has become a major consideration as well. For mass-produced vehicles that are intended for public sale, the additional cost of systems

Automotive System Safety: Critical Considerations for Engineering and Effective Management, First Edition. Joseph D. Miller.

offered to the vehicle manufacturers (VMs) by competing suppliers has a large multiplier. Every dollar may have a six-figure multiplier just for direct material cost. The change in selling price to cover this cost is another multiple and is a significant factor for competitive pricing. Likewise, safety is a major expectation of consumers and is used as a differentiator to win consumer business. To achieve this capability, the industry supported the development of the international standard ISO 26262. This standard requires customers to evaluate suppliers for safety culture, and to consider this in the award of business for safety-related products. Meeting the requirements of this standard requires the expenditure of additional resources. How can these potentially conflicting considerations be satisfied? Both lower cost and evidence of safety are needed to be competitive in the automotive industry. This presents an especially difficult engineering and management challenge. Clever architectures must include clever safety solutions. Systematic architectural analysis supports this.

System Interfaces

In an automotive vehicle, the architecture including the safety-related systems can be quite granular. This architecture encompasses the entire vehicle and may include functional blocks for all the systems needed to provide the required functional concept. This is to ensure that the market requirements are satisfied and also may support make versus buy decisions for the systems providing the needed content. Like any architecture, there are high-level requirements for each of these functional blocks, and these are included in procurement specifications. There is also a communication system for the required intercommunication among the functional blocks. Not only must the vehicle communications structure support the systems necessary to provide the required features, but the interfaces for these systems also must be defined. For example, the steering system may supply the steering angle to the stability control system. The protocol for this communication must be defined, as well as the message content, message frequency, and any safety and security extensions such as a cyclic redundancy check and a message counter for protection against corruption and stuck messages. A stuck steering angle could induce an unwanted stability control intervention.

At a minimum, any conflicting requirements must be resolved. This may require arbitration of requests by the vehicle control systems, such as the steering system. However, to support cost competitiveness, the allocation of requirements must be optimized. Synergies are realized among the systems that reduce the cost, to meet vehicle requirements, including requirements for system safety. A systematic approach is required.

Hazards that may result from the failures of each system can be determined as discussed in Chapter 6. The authority of each actuator is studied to determine whether errors may result in harm to people. The severity of this harm is also evaluated as well as exposure and controllability in order to determine the resulting automotive safety integrity level (ASIL). Then, considering each of the interfaces of each system, the failures can be examined using the hazard and operability study (HAZOP) technique previously described. Errors in content and timing are studied using appropriate guidewords to evaluate the effects of these errors for each message.

Now, as when considering functional safety, a solution is sought, resulting in the minimum allocation of ASILs. ISO 26262 provides for this allocation using the decomposition of requirements with respect to ASIL. This ASIL decomposition is intended to provide a more economical approach to satisfy the requirements of an ASIL. For example, consider the example previously discussed of a lane-keeping assist system consisting of a camera in communication with an electric steering system to provide the correct lane-keeping assistance. The camera observes the lane boundaries and determines the handwheel torque necessary to achieve the correct position of the vehicle between these boundaries. First, consider the allocation where the camera has the requirement to command no more than 3 NM of assist to achieve lane keeping. This is the torque that has been determined to be safely controllable by the general public if applied in error by the steering system. Even if it is applied in the wrong direction, the operator can overcome this level of torque safely in all cases. Now, suppose the camera commands maximum assist when not required. This maximum torque is applied in a direction that causes a lane departure in all driving scenarios, which cannot be controlled by the driver. The resulting ASIL attribute for applying the correct torque is ASIL D. The design to achieve this in the camera incurs all the cost for satisfying the requirements for achieving ASIL D for the system, hardware, and software as well as establishing and maintaining the inherent process with enough independence for assessments and audits. This process may use internal or external auditors and assessors, or a combination, as described in Chapter 2. The steering system must meet this requirement as well – requirements are needlessly redundant.

Next, consider the lane-keeping assist system with the limiter moved to the steering system. The steering examines the signal from the camera to see if the 3 NM command limit has been exceeded. If it has, then the steering system limits the command or ends lane-keeping assistance and informs the driver. The ASIL of the requirement to provide the limit and resulting action is decomposed to ASIL D(D). The burden of meeting the requirements for ASIL D(D) does not result in an additional system safety resource burden for the steering development because the steering system is already meeting ASIL D requirements for providing steering assist in the correct direction, as well as other requirements with an ASIL D attribute. The camera still has the same interface requirement, but the ASIL attribute is reduced. The new ASIL attribute for the camera to provide the correct torque command to the steering system is ASIL QM(D). This enables a significant reduction of resources compared to meeting the requirements for ASIL D: the cost of meeting redundant safety requirements is therefore mitigated by applying ASIL decomposition in this manner. Similar mitigation is applied to longitudinal controls. For example, the engine management system is already meeting requirements with an ASIL D attribute. By applying limits on acceleration commands from the active cruise control system in the engine management system, ASIL decomposition is used again to reduce cost in the cruise control system. The architectural requirements are efficiently distributed, so this safe architecture is competitive.

Internal Interfaces

This same architectural analysis technique can be extended to the safety architecture of each system. Each system is represented by an internal architecture that is granular enough to allow assignment of requirements to the architectural blocks so that ASIL decomposition can be employed. This includes the internal functions of the software, hardware, and separate blocks for the sensors. The system design architecture is analyzed as discussed in Chapter 3; in addition, as the requirements are chosen to mitigate failures at each interface, minimization of the ASIL is a consideration. One opportunity is the interfaces of the sensors with the software. Limitations of the sensor output by the receiving software component may detect out-of-range signals as well as conflicting signals if redundancy is used for detection. This may be accomplished by a software architectural block that is already meeting the higher ASIL requirement, or a block can be assigned for this purpose.

For example, consider a steering system that is receiving torque requests from several sources, each with limitation requirements on the torque that is requested that have been elicited at the VM using the architectural analysis. If these torque-limitation is provided by different architectural blocks of the software that are satisfying only other requirements with no or lower ASIL attributes, then the cost of verification will increase because the other requirements are not sufficiently independent. Therefore, all the requirements inherit the highest ASIL attribute. The developers of the steering system may choose to have a single architectural block that includes these limitations rather than disperse them to the signal-handling functions. This allows the requirements for independence to be met. The cost of meeting the requirements having a lower ASIL attribute is not increased, which helps contain the cost of meeting the safety requirements and therefore resolves the conflicting priority of cost. Likewise, as advised in ISO 26262, higher ASILs are assigned to diagnostic software. Then the functional software is developed to QM or the lower decomposed ASIL requirements if sufficiently independent. Functional requirements verification is less complex for QM or the lower ASIL. The process costs for development and verification of this software can be lower than if it is developed to the higher ASIL. This may lead to lower costs overall and reduce safety process costs.

Requirements Elicitation and Management

Three Sources of Requirements

When eliciting and managing requirements, all potential sources must be examined. The potential sources can be broader that the requirements provided by the customer. There are three potential sources for safety requirements:

- Requirements that flow down from the safety goals and customer-provided system safety requirements. These may be included in the specification from the customer at the time a request for quotation [RfQ] is issued. The customer may include requirements for functional safety as well as for safety of the intended function [SOTIF]. These safety requirements may have been derived from a hazard and risk assessment performed by the customer. Other requirements are derived from a similar hazard and risk assessment performed by the supplier.
- Requirements derived from safety analyses. The supplier may perform safety analyses such as a fault tree analysis [FTA], FMEA, hardware architectural metrics analysis, and software safety analysis that discover requirements for redundancies and other safety mechanisms. This is in addition to many of the requirements that flow down directly to hardware and software from safety goals.
- Requirements derived from assumptions of others, such as interface requirements. These may include requirements for the use of safety elements out of context as well as

assumptions of other systems about the system under consideration in order to meet safety requirements: for example, recognizing and mitigating out-of-bound signal conditions.

As in any industry, automotive suppliers are highly customer focused. Nevertheless, to support completeness, it is not enough to merely satisfy every safety requirement provided by the customer, at the lowest possible cost. Likewise, it is not enough for the customer to ensure compliance with every safety requirement elicited by the customer's safety analyses. Both of these are necessary; neither is sufficient.

As previously discussed, consider an active cruise control system that receives a driver setting, determines the required following distance based on that setting, and sends an acceleration or deceleration command to the engineering management system. The driver setting is for a vehicle speed and is set using the human machine interface (HMI) agreed on with the VM for the vehicle. The following distance depends upon regulations as well as the preference of the operator, also using the HMI agreed on with the VM for this vehicle. These two settings along with the sensor inputs are used to determine acceleration and deceleration commands to the engine management system to maintain this speed and following distance. Customer requirements for the maximum acceleration and deceleration are commanded (as well as regulatory requirements, in some countries). These requirements must be satisfied by the supplier of the cruise control system while also achieving the required control and considering the comfort of the vehicle passengers and operator. They are studied by the supplier of the cruise control system, and an architecture is developed. The architecture is analyzed for potential failures, and safety requirements are derived. These safety requirements include providing a limitation on the acceleration and deceleration to be commanded. After optimization, the supplier determines that to meet the requirement, independent limitation of the acceleration and deceleration commands is necessary in case of failure. This eliminates the possibility that a potential failure that causes the acceleration or deceleration command to be out of limits may also cause a failure in the safety mechanism to limit this out-of-range command. Such an independent limitation could potentially be provided by the cruise control system at some cost or be provided by an external system. This limitation would be most efficiently accomplished in the engine management system, so this is the assumption of the supplier. The engine management system is an external system, it already exists, and it is already provided with the acceleration and deceleration commands of the cruise control system. The supplier communicates this to the customer, and the customer accepts the requirement.

In addition to requirements from the assumptions of others, VM safety requirements are derived from potential hazards that may be caused by vehicle behavior. These are hazards related to vehicle dynamics such as lateral motion and longitudinal motion, which are influenced by the functions of numerous systems of the vehicle directly, or indirectly by the behavior of the driver. For example, sudden acceleration of the vehicle caused by the engine management system may startle the driver and lead to a loss of vehicle control in an emergency. These derived requirements include functional safety requirements as well as requirements related to the SOTIF.

For example, the VM may require the engine management system, steering system, and braking system to switch to a safe state in case of a failure. The safe state is a fail operational state that depends on the level of automation included in the vehicle. SOTIF requirements may relate to the amount of validation required for release of the system if the system falls within the scope, or an extension of the scope of ISP PAS 21448. The VM may also assume responsibility for requirements derived from the vehicle architectural analysis described earlier. These include requirements for intersystem communication as well as safety requirements at the boundaries of the systems included in the architecture.

All requirements elicited from these three sources are captured in a database to support management. This database may include provisions to keep metrics on the status of all the requirements captured, such as their status, ASIL, and the verification status. These metrics help manage the resources that are needed to advance the status of requirements at the rate necessary to support the vehicle program. Traceability to both the source and verification method and status ensures that the required bidirectional traceability can be maintained. Some requirements are considered verified when all their child requirements are verified. Traceability to the verification method allows determination of compliance with the requirements of standards for verification. Changes are managed as well, considering the parent and children relationships of the requirements. This allows consideration of requirements that are affected by the changes. Owners of the requirements affected by the changes are efficiently identified, which supports critical communication with respect to system safety requirements that affect suppliers or other stakeholders. The owners of the requirements can react to the notifications and manage the impact. Critical issues are efficiently communicated and escalated if necessary: such communication is essential for system safety, and the VM can ensure that it takes place.

Cascading Requirements

This elicitation and management of system safety requirements is cascaded to the tier 1 supplier. Management of system safety requirements by tier 1 suppliers is demanded in the automotive industry and a topic in every safety audit. Evidence that the elicitation of requirements is complete and evidence of compliance with these requirements is essential for automotive system safety. The tier 1 supplier also sets up a database of requirements from all three sources for traceability to both the source and verification method and status to ensure that required bidirectional traceability is maintained. Parent requirements include requirements from the VM; requirements that the tier 1 supplier elicited by systematic analyses, such as hardware architectural metrics or software safety analysis; and assumptions of the microprocessor or software operating system.

Changes are managed in the same manner as discussed earlier for the VM. Communication of changes to the requirement owners within the tier 1 supplier or at tier 2 suppliers must be communicated efficiently, transparently, and without exception. Failure to do so may precipitate program delay due to failures in verification, or failures in the field. Assumptions of tier 2 suppliers, such as the assumptions documented in the safety manual for a safety element out of context (e.g. a sensor) are completely captured and traced to verification to support achieving the required ASIL for the sensor signal, including safety mechanisms included external to the sensor. Metrics on the requirements in the database help manage actions to complete this traceability in the allotted time. Additional resources are assigned as required to ensure this achievement.

If necessary, the tier 1 supplier initiates a development interface agreement (DIA) with tier 2 suppliers. This ensures that there is clear communication between the safety managers of the tier 1 supplier as the customer and the tier 2 supplier. The relationship may also be beneficial for the management of SOTIF because of the dependency of SOTIF verification planning on the internal processing of the sensor. While this internal processing is confidential and proprietary to the tier 2 supplier, it must be examined to ensure the safety of the system. Therefore, including requirements for the exchange of information concerning SOTIF requirements in the DIA supports achieving system safety. While this is sensitive information to both parties, a degree of understanding of tier 2 product design sensitivities is necessary by the tier 1 supplier to avoid inadvertently triggering malfunctioning behavior in the algorithms due to these sensitivities, as well as to understand the limits that need to be verified. This is accomplished by reaching an agreement with the supplier so essential information is not withheld, and the confidentiality and intellectual property considerations of the tier 2 supplier are respected.

Understanding the tier 2 supplier's logic is also useful in order to analyze and hunt triggering events. Measures are taken by the tier 1 supplier to compensate for limitations and ensure that the intended function is safe when these limitations are exceeded. This compensation supports all the areas of SOTIF verification and validation. Performing analyses at the limits, exercising known scenarios at these limits, and emphasizing exercising the system where potential unknown unsafe scenarios are encountered are all supported. The tier 1 supplier's assumptions of others are communicated as requirements to the owners of the other systems involved: these may include support such as information from additional redundant sensors to compensate for sensor failures, and takeover by redundant directional control systems in case of actuator failures. These assumptions are verified and managed, which supports traceability.

Conflicts with Cybersecurity

Conflicts may arise between requirements for system safety and requirements for cybersecurity. ISO 26262 requires communication between the domains of cybersecurity and functional safety to prevent the neglect of information elicited by each that may be useful to the other. Both are competing for the scarce resource of throughput bandwidth; processing security measures and safety measures can be computationally intensive. The detection period for safety mechanisms must be respected in order to meet the fault-tolerant time interval without causing false detection.

Consider the case of communication in a failsafe system that contains a hash code for data protection. This hash code is evaluated to ensure the security of the associated signal; failure of the hash code to be confirmed indicates a breach of security. Hardware is dedicated to checking for corruption, and if corruption is detected, a predetermined security action is taken. This security action must be agreed on with the VM and not adversely affect system safety. The hardware is included in the computation of architectural metrics of the system hardware. In order to meet functional safety requirements, this hardware may need diagnostic coverage to determine whether the computation is completed correctly, or if a failure could lead to a violation of a safety goal.

When such diagnostics are included, a safe state must be identified that will disable this security protection. There are many systems in the field without such security protection, so this is not seen as an unreasonable risk – the system may continue to operate safely without it. On the other hand, it may be decided that if a failure is detected in the hardware for decoding the hash code, the safe state is to disable the system. Otherwise the system is unprotected against a cyber-attack, and this vulnerability can be exploited. This conflict needs to be agreed on at the supplier and VM level.

The VM may have an overall security strategy for the vehicle that has been agreed on with the safety specialists and the VM executives and that is to be deployed consistently to all safety-related systems on the vehicle. The supplier may suggest alternatives and the advantages and disadvantages if such a mandate is not being deployed.

Consider the same issue in a fail operational system, as may be required for some levels of autonomous driving. The level of risk of a cyber-attack may be considered greater, so an alternative to disabling the hash-code check is required even if the hardware checker fails. Now, if the fail-safe choice is made, a redundancy is needed to achieve the fail operational state, or a minimum risk condition. This fail operational state may not have the hash-code security or can contain a redundant means to retain this security. A strategy is also needed for how long to maintain this condition as well as how to end it. These choices need to be identified and made at the system level. Stakeholders of both safety and security should be involved to determine and evaluate the alternatives, as well as stakeholders from the engineering domains of hardware, software, and systems.

It is advisable to assign a specific task to system engineering to identify and resolve conflicts between system safety and cyber security, such as establishing a task force for this purpose. These may include conflicts not only about availability, but also regarding computational time and memory. Such known conflicts may exist because safety and security features have not been reconciled in components such as microprocessors and other integrated circuits before they were released for sale. While they are challenging to resolve, resolution is achieved by a holistic approach in the system. The requirements flow starts with vehicle features. Then an architecture is developed and analyzed to determine requirements for each supplier. System engineering at each supplier reconciles the safety and security requirements, and an architectural analysis can then determine the safety requirements for the tier 2 suppliers. There is feedback at each step, and the task force can diligently use this flow. Figure 8.1 illustrates the requirements flow.

Determination of Timing Risks in an Automotive Application

Milestones

Automotive timing for a launch-critical program that targets a model year is obtained through diligent program management. This program management is cascaded to all functions, suppliers, and domains of engineering and operations supporting the program. As the program management tasks are cascaded, the tasks and milestones become more granular. The overall program timing is fixed and supported by these cascaded program plans. Delaying a planned vehicle launch can have significant economic penalties for the VM as well as the automotive suppliers: not only is the cash flow less favorable with a delay, but overall sales are reduced due to early sales being lost to competing products.

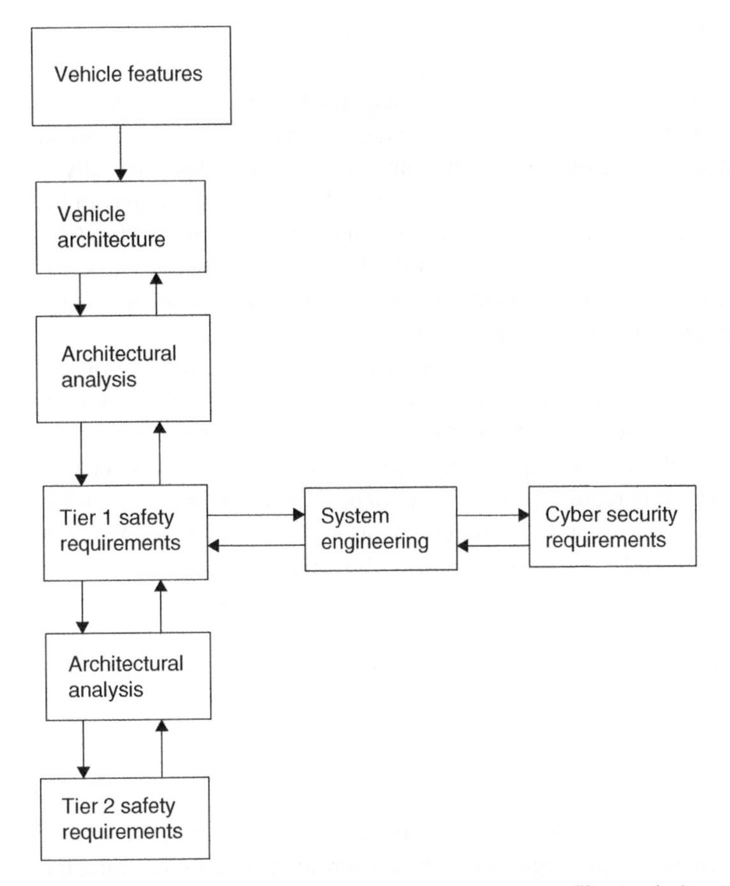

Figure 8.1 Requirements management and security conflict resolution.

Meeting VM milestones has a very high priority for everyone contributing to the launch program. Each VM has its own names for milestones, and these names are embedded in the internal process for the VM and are consistent for every vehicle launched. The internal names are used with suppliers to promote consistency in communications with the VM, and a different VM may give similar milestones to the same supplier with different names.

Samples

While these milestones are granular and varied, in general, all VMs have milestones for concept samples, design verification samples, and production validation samples. The milestones refer to samples because these samples generally require deliverables from all suppliers on the program, whether internal or external. Each sample is used for essentially the same purpose by all VMs at their respective lifecycle phase, such as concept verification. These are sometimes generically referred to as *A samples*, *B samples*, and *C samples*, respectively. Such generic references provide clarity of communication in forums where multiple VMs and suppliers are communicating, such as during standards development. Then any applicable standards can be easily adopted to the VM's process and used consistently, such as ISO 26262.

It is not unusual for A samples and B samples to be scheduled earlier in the program than expected in order to reduce schedule risk later in the program. While doing this is resource intensive, the launch schedule has high priority and may justify the additional expense. This expense is incurred at both the VM and the suppliers supporting the VM for these samples. It is also not unusual for resources to be ramping up on the program, especially at suppliers, as quickly as possible after award of the program. Moving the A sample and B sample forward will cause this ramp-up to be even steeper and harder to achieve. The effect is multiplied at the VM and suppliers if there are multiple programs targeting the same launch window. Because of this, there is competition for scarce resources in order to complete the concept and design system safety tasks on time.

Many tasks fall in the requirements development area and therefore require a resource-intensive effort by systems engineering and system safety engineering. This is followed by an equally intensive effort with respect to requirements engineering in the software and hardware domains prior to these milestones. Completing these tasks is critical because they determine the system safety requirements that are to be included in the design for B samples and then verified. These are not only requirements to be verified by the VM, but also requirements to be verified by the supplier during this same period of time on the program. Some requirements are verified at the system level, while others are verified at the hardware and software levels. If any of these requirements are not identified, verification is delayed, and this delay could be critical to the program. Complete verification may not be possible.

Program Management

To ensure that system safety tasks receive sufficient priority, it is important for system safety milestones to be included in the program schedule. There are critical deliverables for system safety in every lifecycle phase of the program leading up to program launch. All of these require resources that are also needed for tasks relating to each of the program samples. In Chapter 2, ownership of system safety by program management was identified as one of the five critical considerations for success of the system safety organization. Resolving resource conflicts while the needed resources for the program are being ramped up can be a difficult program management responsibility. Nevertheless, unless it is resolved successfully, system safety may fail to be included on the program consistently with the required enterprise system safety process as needed for compliance with the enterprise system safety policy.

One way to demonstrate ownership by program management is for the program manager to include system safety milestones in the schedule, as well as the tasks and owner of the tasks. This provides the needed transparency of progress for system safety compliance as well as clarity of responsibilities as required for a positive safety culture. At every program review, this progress can be reported and actions can be determined to maintain required progress on system safety. Such progress can be measured in a manner similar to measuring progress toward the completion of the A samples and B samples. The priority can be made visible and can be reinforced. The program manager can track and report progress directly. Inclusion in the critical path for the B samples is appropriate, and such a priority can be supported by the enterprise. The enterprise system safety policy can mandate such prioritization by requiring compliance with the policy. This behavior is reinforced by inclusion of critical metrics for the program that are reviewed by executive management, as discussed in Chapter 5. This reinforcement is made visible to the program team by the program manager, which empowers the team to comply with the system safety policy and process. Safety audits can supply further support. The program manager is invited to these audits.

During the safety audit, the auditor reviews the system safety process being followed and the development of requirements. Every auditor may emphasize and devote audit time to the review of requirements, because system safety requirements development is so critical to the successful execution of a system safety process. Unless requirements development is complete and timely, the system safety process will fail. This is a more detailed review than is practical for the program manager to perform, as are other engineering reviews. Capturing the three sources of requirements must be checked, as well as the correct linking of parent to child requirements. Adequacy of the requirements database and included attributes must be checked. Traceability to verification as well as the method of ensuring that all the verifications are complete and satisfactorily ensure compliance must also be checked. However, if there are improvements to be made in order to support the launch schedule, the system safety auditor will recognize this very quickly and perhaps schedule a separate review of a recovery plan and progress.

While the auditor is independent and is not directly affected by the success of the program, the auditor is genuinely motivated to support the successful execution of the system safety program in order to ensure the absence of unreasonable risk to the general public. The program manager is kept briefed on progress and informed about any actions needed, or escalations, before the metric reviews. This prepares the program manager to be successful in these reviews, and to advocate for the needed considerations to execute the system safety program. Since this is supported by the enterprise policy and the metrics of the enterprise, executive support is expected. Such a symbiotic relationship supports success of the program and therefore supports the program manager. System safety metrics on the program are more favorably reported to senior executive management.

Design and Verification

Sample Evaluation

As discussed earlier, significant priority is placed on the design team to achieve the dates for the A samples and B samples – the economic and market motivations are too compelling to resist. These priorities are supported by the entire supply chain of the automotive industry. To meet the date for the A samples, the essential functional requirements need to be understood and included in the design of the A samples so that their impact on the concept is evaluated. Limitations on actuator authority or limitations in intended functionality need to be evaluated and reconciled prior to the final determination of the design content The A samples, unlike the B samples, in general will not be exposed to the general public, so safety requirements intended for production are not essential for this phase of the program. The remaining requirements will be essential when the B sample design is complete.

Which requirements are essential for the A samples must be determined; these requirements are needed for the concept evaluation and to ensure that the risk is acceptable for the use of the A samples. It is important to note that the risk considers not only the A sample operator, but also the other users of the evaluation area. It is appropriate to formally determine the use of the A samples in order to ensure that they are fit for purpose. For example, an A sample can be used to evaluate a potential hazard to determine whether a legacy safety mechanism is adequate for a new application. In this case, test software is needed to induce the fault that is to be mitigated. The safety of this test must be formally evaluated and documented to ensure not only that the verification method is adequate, but also that the test method is safe for everyone involved or exposed to the test. Use of an A sample on public roads may require that any potential failure is controllable by the operator or by the system in a manner that does not expose the general public to an unreasonable risk. Such evaluation of controllability must be accomplished in a safe environment that does not expose the general public to this risk, such as a testing area at a test ground with established safeguards in place. After such controllability has been evaluated, reported, and reviewed as sufficient for public road use, even under failed conditions, then public road use is approved.

For each stage of use of both the A samples and B samples, a progression such as this is employed to determine fitness for purpose. If there are changes to the samples or the intended purpose, then a formal safety review is required before further public road testing is approved. Such formality is needed to ensure that there is evidence of diligence before exposing the general public to this revised evaluation. For example, safety-related systems for the A sample may be verified in a lab environment prior to being used in a test vehicle. Faults are inserted on a test bench to evaluate whether they are controlled as well as for the safety of the system behavior at the operator interface. This environment may evaluate the safety requirements that are appropriate for the intended use.

Consider the motion of the steering wheel if it is tested on a low mu surface on a test area and a fault occurs that causes rapid spinning of the wheel. This could be hazardous due to the spokes of the wheel. Mitigation is needed for this fault to protect the operator. Now, if this safety system is used on a test vehicle, the operator will be protected. If hazard testing is performed to validate the concept safety analysis independently of legacy safety mechanisms, the conditions need to be studied so that safety is ensured without safety mechanisms. This may not allow some conditions, such as an inserted unmitigated fault like a steering fault, while driving on a high-speed oval track. The fault is inserted at varying speeds in a more open area with acceptable test conditions. Controllability by the test driver of the various systems that are to be tested can then be evaluated in a low-risk condition by the test driver under intended and failed conditions prior to evaluation on the public road. These tests need to include the variety of scenarios encountered in public road evaluations. Special instructions can be provided to the test driver to help obtain this controllability, if necessary. Such evaluation is formally planned and reviewed before it is executed, and a pass or fail criteria is determined in advance. This is especially important for systems with higher levels of automation, as the pass-fail criteria are unique to the worst-case scenarios that are anticipated. Objective criteria are collected in different test conditions that are evaluated by these worst-case scenario criteria. After the tests are run, a report is prepared and reviewed as discussed previously. This provides evidence supporting the intended use and supports diligence.

Verification

Verification in an automotive program is intended to mean verification of requirements. Sometimes, the term *verification* is used in other contexts to refer to tests, such as environmental or electrical tests. These tests are run on systems for many programs to verify requirements that are common to them. Nevertheless, verification must confirm compliance with all requirements, not only common requirements.

The need for verification is driven by the need to meet the requirements that the VM has determined are appropriate for the features included in the vehicle model. These requirements may be unique to the features as modified for the intended content for a particular feature, and may include both functional requirements and system safety requirements. It is important that a complete set of requirements is elicited for verification. As discussed earlier in this chapter and illustrated in Figure 8.1, these requirements come from three sources: they may flow down, come from safety analyses, or be assumptions of others. Requirements that flow down may be allocated from customer requirements to the system architecture as well as requirements for hardware and software. Requirements that are elicited from safety analyses come from safety analyses of the system, software, and hardware. And requirements derived from the assumptions of others may come from interfacing systems, hardware SEooC, and software SEooC, such as safe operating systems.

Much of the verification may be achieved by testing. This testing, in order to be traceable to individual requirements, must have test cases traceable to the specific requirement being verified. The pass-fail criteria must be appropriate for each requirement. Since the requirements that flow down may be repetitive from program to program, a standard set of tests for design verification is often developed to support scheduling efficiency for multiple programs, as discussed previously. These tests are appropriate for the common requirements as long as traceability is established for the requirements of each program. When an audit is performed on an individual program, it should include checks to ensure that safety requirements that are verified in this way are complete, including all three sources; that they are traceable to the individual test case that verifies each of them; and that they all pass. Findings may include gaps in the ability to confirm which individual test case verifies a particular requirement, or which requirement a test case is verifying. This bidirectional traceability is required for compliance with ISO 26262. The evidence is important for the safety case and supports diligence.

9

Other Discussion and Disclaimer

Background

Among the reasons that the automotive industry expends significant resources to assure the achievement of system safety for products is to reduce the risk of safety-related recalls. A single recall can cost hundreds of millions or even billions of dollars. This is just the direct cost, such as replacing ignition switches because of potential fire issues on 28 million cars at a cost of $20 each. The direct cost of some recalls concerning accelerator pedals or airbag inflators has been estimated in the billions of dollars. In addition, sales may be lost because of lost goodwill. The ISO 26262 standard makes no claim that compliance will avoid safety-related recalls; it is a compilation of the judgment of industry experts over many years of what is required to achieve functional safety, and what artifacts are to be assessed to assure compliance with these requirements.

Following the guidance in the ISO 26262 standard to compile a safety case is intended to show evidence of the safety of the subject item. A safety case for functional safety is a compilation of the work products to show evidence of compliance with ISO 26262 and the argument that the item is safe. However, if the product fails safely too often, there is a recall. The safety case may contain evidence that this should not happen due to random failures.

Likewise, ISO PAS 21448 does not claim that if its guidance is followed, there will be no recalls related to the safety of the intended function (SOTIF). A safety case that includes compliance with ISO 21448 may provide evidence that the likelihood of a safety-related recall based on the safety of the item in normal operation is not unreasonably high. Because of the "dread" factor of public reaction to a safety-related issue with an automotive system associated with the levels of automated driving, the risk of an issue is shown to be quite low. No guarantee is provided that compliance will prevent recalls; ISO PAS 21448 is a compilation of the judgment of industry experts over many years of what is required to achieve SOTIF.

Work was started on the safety of a system without a fault as early as 2014 with a small international group of suppliers and VMs. The relevance of this aspect of system safety became more apparent as advanced driver assistance systems contained more automated features. Using both these standards to improve system safety is well advised.

Automotive System Safety: Critical Considerations for Engineering and Effective Management, First Edition. Joseph D. Miller.

Consider an automated emergency braking system. The application of ISO 26262 should provide significant evidence that emergency braking will not be induced by the failure of a component or the software in the system. It is expected to be reliable. Validation in accordance with ISO PAS 21448 may show that the risk of the automated emergency braking system causing a rear-end collision is 10 times less than the risk of a rear-end collision in the field. Requirements for software verification exist in ISO 26262. The software in the field is expected to be compliant to these requirements. Compliance with automotive software quality standards may also be expected, and maturity of the software process is demonstrated. There are requirements for diagnostic coverage and random failure rates. These are normally met when compliance is shown with ISO 26262. The requirements for safety mechanisms are elicited while the hardware architectural metrics are computed. Requirements are audited for compliance and traceability; these numeric metrics may attract attention, and their importance may grow.

Three Causes of Automotive Safety Recalls – Never "Random" Failures

Failure Rates

During hardware development, different sources of information are used to predict random failure rates for components and their allocation to failure modes. Much of this information is not from suppliers or from data compiled recently for such failure rates in in automotive systems. They also may not be historical data from the producer of the system based on actual field data from previously launched systems. Some sources are standard for that purpose and are referenced in ISO 26262; other methods to estimate random failure rates can be used in the metric calculations for different automotive safety integrity level (ASILs).

Since the single-point fault metric and the latent fault metric are ratios, the absolute value of the failure rate used is not as influential for determining diagnostic coverage of hardware faults. The exception may be if there is an error in the distribution of faults if one of the failure modes is not well covered. For example, consider a thermistor for which the most likely failure mode listed is a short circuit. Since it is a pressed powder part with end caps manufactured using a vibrating bowl feeder, the most likely failure is an open that is induced mechanically. The erroneous failure distribution in the data source may completely reverse the coverage of this part.

Nevertheless, metric calculations are useful for eliciting requirements for diagnostics. They provide a systematic method of considering each failure mode of every component, focusing only on the safety-related effect. Then the required safety mechanism for each failure mode is examined, emphasizing the effectiveness of the safety mechanism to cover each of these failure modes. Even if the rates are significantly in error, eliciting the diagnostic requirements with this level of rigor is important. The requirements are specific enough to be developed and implemented directly. They are verified directly, which ensures the coverage that is needed for system hardware regardless of actual failure rates.

The overall probabilistic metric for random hardware failures (PMHF) is more influenced by the source used to determine random failure rates. In this case, the metric is not a ratio; the failure rates of the components are additive. There are no mitigation errors built into the standards employed unless the confidence is understood and scaled back to 70%, as allowed in ISO 26262. Otherwise, the overstated failure rates may cause additional cost that can make safety-enhancing systems less affordable for the general public. Substantial effort is expended to meet this number using sources known to be extremely conservative. Some of these efforts include redundant components that would not be required if the failure rates used were less conservative. While the complexity may grow, the PMHF is more favorable. The motivation is to reduce recalls, and this is a well-intentioned effort.

Recalls Due to Random Hardware Failures

Over many years of reviewing automotive recalls posted by the National Highway Traffic Safety Administration (NHTSA), one will observe few, if any, recalls due to random hardware failures. Certainly there have been many recalls due to hardware failures: the example used earlier was a hardware failure; the ignition switch overheating was a hardware failure; there have been recalls of braking systems due to hardware failures; and there have been steering recalls due to hardware failures both electrical and mechanical, such as intermittent connections. There could be hardware failures such as an avalanche diode that overheats, causing a fire; a microprocessor that experiences failure in its random-access memory; or a ball joint that fails under load. But none of these are *random* hardware failures. The random failures of parts in automotive applications is never high enough to cause a recall by itself – there must be a systematic cause for the failures to occur over a short enough period of time to exceed the still relatively low threshold for a recall:

- An avalanche diode overheating may result from a bad lot of parts from the diode supplier. The supplier may have had an internal quality spill due to a process going out of limits and the parts somehow inadvertently being included in a shipment. The failure rate for that lot is much higher than the overall random failure rate for the part, and the random failure rate may still be much lower than predicted by a standard. A recall is caused by the short-term spike of failures.
- Failures of random-access memory may result from a processing defect that developed a semiconductor manufacturing site that was not detected by the quality system at that site. Again, the failure rate for that manufacturing site is much higher than the overall random failure rate for the part during the period of time the processing defect was undetected. Still, the overall random failure rate is much lower than the conservative prediction of a standard. Again, a recall could be caused by the short-term spike of failures.
- A ball joint failure may result from a forming error at factory that escaped the quality system at that factory, or a design error in determining the load capability. Neither cause results in a random failure of the ball joint. The forming error is a machine setup issue. It may have been periodic in time and missed the quality audits systematically. The failure rate for that factory is higher than the random rate for ball joints during this period of time, but the calculated rate is correct over an extended period.

Examining random failures would not prevent these recalls, because the recalls are not due to random failures. The cause lies elsewhere.

Causes of Recalls

To prevent safety-related recalls, the focus is on preventing three observed causes. Mitigating these three causes can prevent or reduce the severity of most safety recalls. The vast majority of safety-related recalls are traceable to one or more of the following:

1) Missing safety requirements, including those that were never elicited from the three sources of safety requirements from Chapter 4, or that were elicited inadequately
2) Requirements that were not verified, including those that were never verified from the three sources of safety requirements from Chapter 4, or that were verified inadequately
3) Quality spills at suppliers, including products released by the supplier to the customer that do not meet requirements initially or in service

For the avalanche diode that overheats causing a fire, requirements to prevent fire and smoke need to be elicited and then verified, including requirements for preventing sources of ignition and preventing flammable material from being exposed to a source of ignition. Leakage of the avalanche diode, for example, must be prevented or detected, or the heat should be detected. The avalanche diode is a source of ignition when it leaks, so requirements for detecting the leakage and removing power from the avalanche diode can remove this source of ignition. Also, if heat is detected, requirements to reach a safe state with a warning are needed. Such a safe state may also include power removal from the system as well as notification to the operator of the vehicle. A requirement to prevent any excessive heat from starting a fire externally is also needed.

Even with these requirements for reaching a safe state and preventing a potential fire, the loss of availability of a safety system due to a quality spill at a supplier, such as a bad lot with too many leaky diodes, could still result in a recall. This could require some early containment at the diode supplier so that the number of failures in the field is not too high. For example, screening or lot sampling for this particular defect as a critical characteristic can minimize the probability of such a defective lot to a level that should effectively prevent a safety-related recall. The controls reduce the number of defective parts under a worst-case condition. The requirements for countermeasures such as those described against fire in the system help reduce the potential severity.

All these requirements need to be verified. Verification of the countermeasures may include fault insertion or simulation with criteria for maximum external temperature and observation of no fire or excessive smoke. Surveillance of the supplier's management of critical characteristics can be used for verification and quality control. Such failures may fall outside the scope of ISO 26262. ISO 26262 considers only functional failures, so while the diode is included in the architectural metrics, the diagnostic coverage may not include countermeasures effective against fire and smoke.

Now, consider the failures of random-access memory discussed earlier. These failures may cause values in memory to change unexpectedly. These values could be safety-critical dynamic variables, and an unexpected change might cause an unexpected unsafe behavior of the vehicle. If the requirements for detecting such failures and either going to a safe state or failing operational with a warning are identified and verified, the potential safety-related recall can be avoided, although warranty returns may increase.

For example, a systematic software safety analysis at the system level may identify errors that are caused by the exception of a hardware error. The cause of random-access memory failure is identified, or the error is handled at a higher level, agnostic of cause. A more-detailed software safety analysis may identify the specific critical variables to be protected. Either way, verification is necessary to ensure that the protection functions correctly.

This also requires some early containment at the fabrication facility so that the number of failures in the field is not too high. Screening or sampling at the fabrication facility to test that such failures will not happen, perhaps using an accelerated stress test of the memory, is needed to ensure that the probability of shipping enough defective parts to cause a recall is low enough.

Also, for example, consider an electronic stability control system that malfunctions under some conditions that are not related to a hardware failure, and that causes a recall. This is different from the quality spills just described, because it is not traceable to a manufacturing defect. No hardware component failed to meet its specification when released for production. Such a failure is assumed to be caused by a software failure, or "bug." The assumption is that since the hardware functions correctly and met its specifications, it must be a problem with the software not meeting its specifications. However, since software does not experience random failures, the failure must be systematic, as are all software failures. The malfunctioning behavior that is observed can be traced to the elicitation, compliant execution, and verification of requirements. Either the requirements not to exhibit the malfunction were missing or inadequate, or they were not verified or verified inadequately. There is no other plausible explanation. Even if the software "learns" and therefore changes over time, adequate requirements elicitation and verification must take such learning: into account. Had these requirements been verified, the malfunctions would not have occurred, and the recall would have been prevented.

Completeness of Requirements

Early identification of requirements with confidence in their completeness is necessary to minimize recalls. The early identification of requirements supports developing a compliant concept and a compliant design. As more details of the design emerge, additional expansion and refinement of the system safety requirements is possible, so the requirements consider the entire design.

Confidence in completeness comes from systematic analyses of hazards within the scope of the standards and even of hazards not covered by the standards, such as avoiding fire and smoke. These systematic analyses can be expanded hierarchically to ensure that the granularity is sufficient. All three sources of requirements are examined, as discussed earlier in this chapter. This confidence in the identification of risk was discussed in Chapter 7 and includes consideration of the actuator authority, communication with other systems, fire, smoke, and toxicity. Requirements to mitigate these hazards are systematically identified. A rigorous systematic consideration of all the use cases is important for both functional safety and SOTIF.

ISO PAS 21448 provides some guidance for eliciting relevant use cases in order to evaluate potential triggering events. This is further discussed in Chapter 3, including arguments for completeness. All these analyses should ideally be finished before the design is completed. Then there is still the possibility of modifying the design to obtain compliance with the requirements prior to the start of design verification. This supports completing the launch schedule on time, including B samples that are used on public roads. Risk is added to the project schedule if the analyses are completed while design verification is in process.

Timing Risk

Then the design and the verification test plan must be updated. A managed recovery plan is needed to support completing this design update and verification in time for validation and the delivery of C samples that are off production intent tooling. Further risk is added if the analyses are later – for example, during product validation. Any new requirement discovered may require a modification to the design and repetition of verification. The redesign may also require modification of the validation plan and repetition of some of the validation activities. This could be especially significant with respect to SOTIF validation: it may require modification and repetition of the simulations used, as well as road testing. Such repetition may precipitate further design modification if unknown unsafe scenarios are discovered.

Should the analyses slip past project release, then the risk of a potential safety-related recall may increase. Such analyses may discover a safety requirement that was missed, and there is no opportunity to correct it in the field without recalling the product. The potential for harm may exist after the recall if the recall is not complete.

Likewise, requirements for manufacturing to detect and contain any safety-related defect are needed. Critical characteristics need to be managed, as well as any testing for safety-related functionality, including surveillance of suppliers. Evidence of compliance with safety requirements for production is necessary to detect and contain safety-related quality defects. The role of product quality is key: quality assurance helps prevent recalls.

"But It's Not in the 'Standard'"

An experienced auditor may say during an audit, "It is advisable for one to comply with applicable standards, and it is necessary to also make a safe product." The auditor may tell stories based on personal knowledge or experience to make the point clear that safety-related field issues can result from causes outside the scope of ISO 26262. This prior knowledge may make the case even more compelling to be diligent to address such causes and include the evidence in the safety case. The safety policy and safety process of the enterprise may support such diligence. The auditor may escalate issues that are not in the scope of ISO 26262. Too often the auditor may hear, "Why must we perform fire and smoke testing? It is not required by ISO 26262. The customer only requires meeting flammability requirements." Or "Why should I care about toxicity? It's not in the standard."

With the schedule and financial pressures in the automotive industry, and the wide endorsement of automotive functional safety according to ISO 26262, these demands are in focus for automotive product developers. Other aspects of system safety may not have the same visibility. Conferences and certification classes may regularly discuss ISO 26262, while other aspects of system safety have no external stimulus to promote such careful consideration. When seen in the context of the preceding chapters, such questions may seem naïve or even shocking. Harm to people from any automotive system related to any cause must be diligently prevented, and everyone contributing to product development and production shares in the responsibility to prevent unreasonable risk to the general public. However, if a customer hires an outside auditor to check for compliance with ISO 26262, then an automotive supplier may emphasize this priority over other system safety priorities to satisfy the short-term need for a favorable audit.

Competing Priorities

Such prioritization is compelling when taken in the context of the other competing demands of the launch program. The result may be reduced availability of resources and time demands for other safety requests not related to compliance of the standard being audited. Management may become sensitized to this compliance, driving scheduling and resource decisions to meet requirements of the "standard." Findings by the customer's ISO 26262 auditor may further escalate this sensitivity and be used to justify the necessary resource assignments to resolve the findings. These actions support diligence to ensure functional safety of the product – this is the result intended by the audit. Then, when additional requirements for system safety are elicited internally that require further schedule and resource prioritization for compliance, there may be pushback from internal resource groups because of the conflicting priorities set by the customer and their internal management. The supplier will be driven by the customer's priority and will make satisfying this priority with respect to functional safety an actionable goal supported by the supplier's management. This support will be visible in the related tasks resourced and executed in the program plan and will continue to ensure that the ISO 26262 auditor's findings are resolved. In this context, the questions and pushback seem perfectly reasonable.

How can these other internal findings be given equal priority to the customer's and internal management's prioritization of compliance with ISO 26262? Current and future business is seen to depend more on the customer's safety priorities than on other internally generated safety requirements. From the perspective of a demonstration of diligence, evidence of such resistance is not helpful. If a field problem is being investigated, evidence of pushback may not appear supportive of a favorable safety culture. This may outweigh the constructive actions taken to ensure functional safety. Support for all safety requirements supports diligence. It is better to preclude such questions, and doing so requires prior goal alignment.

As is normal for achieving a successful joint venture, alliance, merger, or other competitive alignment, it is imperative to understand and ensure that the direction taken satisfies the goals of all stakeholders. This is a fundamental tenet of achieving win-win solutions in negotiations to complete such competitive alignments as well as to resolve internal resource conflicts. Without a win-win solution, the alignment is in jeopardy of future misalignment and renewed conflict, and the goals of all parties may not be achieved. Achieving a win-win solution will support the talents and energies of each participant to progress in the agreed direction and will help all to get closer to achieving their individual goals.

A reward system that benefits individuals for achieving these goals is beneficial. It reinforces the importance of each individual in achieving these goals and emphasizes the priority of these goals to the enterprise. Self-interest and greed may help fuel the effort and maintain the momentum. A successful enterprise reward system institutionalizes the rewarded behaviors, such as those needed to ensure the safety of the products produced, and thereby protects the enterprise with evidence of diligence and a positive safety culture. The establishment of common individual goals, rewarded and reinforced by the enterprise, provides empowerment for all stakeholders to resolve potential conflicts.

To avoid such conflicts, as described earlier with respect to achieving a safe product, preemptive steps can be taken to agree on the system safety process. Having a standard process that is agreed on at the enterprise level and that therefore includes all the stakeholders involved in every program of the enterprise supports achieving a win-win solution to potential conflicts. Each stakeholder can strive to achieve compliance and earn the rewards provided by the enterprise for achieving the safety objectives. All the requirements of ISO 26262 are included in the standard process. The required work product content is included and merged with content required by a quality standard, where applicable, to further reduce potential resource conflicts. This process permits a high tailoring of content that is approved with evidence of reuse. Such reuse can favorably affect both the resources consumed to ensure the safety of the product and the development costs for the product itself. This may further enhance the achievement of enterprise cost objectives by the program manager and program participants. Following such a process assures that a customer-directed audit of compliance with ISO 26262 will be brief and favorable, paving the way for new safety-related product business and rewards for the internal participants.

Audits and Assessments

Audits and assessments by internal personnel are included in the enterprise safety process and provide early information to ensure compliance with the process. The internal assessment and audit reports are shared with the customer as evidence of process compliance and used to aid, or in lieu of, external audits. Likewise, the inclusion of nonfunctional safety requirements for system safety in the standard process, such as fire and smoke, toxicity, and SOTIF, allows standardization of compliance measures and minimizes resource demands. Chapter 3 discusses the efficiencies achieved by combining SOTIF with automotive functional safety; these efficiencies can be ensured by compliance with the system safety process.

Work products may have templates that are tools to facilitate the merger of ISO 26262 and SOTIF compliance. The assessment criteria may also include the same ISO 26262 and SOTIF compliance requirements so that favorable assessments are facilitated. Inclusion of fire, smoke, toxicity, and SOTIF demonstrates a differentiating positive safety capability and culture that promotes consideration for more advanced safety-related business opportunities. The risk to a vehicle manufacturer (VM) is perceived to be reduced by purchasing products from a supplier that has implemented such a comprehensive safety process; this perception is further reinforced by observing the enterprise's support for the safety policy and process, as well as a reward system that promotes a positive safety culture. This paves the way for new safety-related product business and rewards for the internal participants. The safety criteria for awards are satisfied competitively by the enterprise because of the comprehensive safety process and evidence of compliance.

Further, the enterprise safety process is scalable and may easily absorb new business with some efficiencies of scale due to the tailoring that results from the increased opportunities for reuse. When this process is tailored and proposed for inclusion in the project planning with the appropriate milestones and responsibilities assigned, it is supported favorably by the other stakeholders because of the potential to advance their own individual self-interest. Achieving each milestone provides evidence of process compliance that is used in internal reviews and reviews with customers. This achievement also provides evidence for the participants to use to justify individual rewards. The inclusion is perceived by executives to reduce product safety risk, and this perception is justified by the evidence produced by the process as well as assessment and audit reports. On-time performance is rewarded, and all stakeholders win.

Disclaimer and Motivation for Continuous Improvement

A spokesperson for system safety might be asked, "Will following the safety standards ensure that a producer will not be sued, or if sued, will win?" "No, but it may provide supporting evidence of diligence," the spokesperson replies. The spokesperson may further be asked, "Then what will ensure that a producer will not be sued?" and reply, "Beats me!"

While this could be considered trite or flippant, the questions and comments are relevant. A safety case provides evidence of diligence to achieve safety of the product and an argument that relies on this evidence for that product's safety. Each person who compiles a safety case, reviews it, or contributes to establishing the policy and process to produce it will contemplate its potential uses. In addition to ensuring diligence for achieving the safety of the product, the safety case may be used to defend the enterprise using this evidence of diligence. Nevertheless, a successful defense cannot be ensured.

Likewise, following every recommendation in this book may not ensure that there will be no defects, no lawsuits, and no damages claimed due to the associated product. These recommendations are the opinion of the author based on his experience. There is no implication that the recommendations are complete, or that they perfectly apply to the reader's situation in every case. The author's experience motivated him to provide such recommendations to help every reader of this book by providing considerations that the author has found to be relevant over the years, that may be relevant for others, and that are not included in the referenced standards. Each producer is responsible for the safety of the products produced. It is easier for the producer to deal with information from the experience of others than to deal with the lack of such information. This book may shorten the period needed to acquire such information from experience; the considerations here are useful in satisfying this responsibility for producing safe products. Many of the recommendations indicate how to get it done in a real enterprise, rather than teaching the content of the standards, because information about the standards is readily available. The producer may consider the advice in this book and choose a direction, and the producer is then responsible for that direction. The discussions here may help.

The direction chosen by the producer may include a choice of system safety organization, policy, and process. Such a choice provides evidence that the enterprise is focusing on system safety rather than using an ad hoc approach where each sub-organization determines policy and process, as well as whether to have a safety organization and how it is structured. The safety policy, process, and organization are interrelated to function in a manner that achieves the safety goals established by the enterprise.

Policy Statement

The enterprise can state its agreed-on safety goals in a brief policy statement in addition to a released policy document; this policy statement should be widely visible to ensure efficient communication. Simple statements that are widely distributed about topics as sensitive as safety and quality may require significant review and approval by executives responsible for determining the direction of the enterprise as well as executives responsible for steering the enterprise in the stated direction. For example, the enterprise may have a policy statement that says its process will achieve safe products. Such a statement can be easily distributed for visibility across the enterprise and will be considered noncontroversial even by the law department. By stating this goal briefly and concisely, it is understood and provides support for efforts by the enterprise to realize the safety goal.

There may be similar policy statements with respect to quality. These statements have a similar purpose: they do not establish the detailed implementation of a strategy to achieve the goal, and they do not substitute for a detailed policy that clarifies objectives related to the goal – they simply state the goal. Such statements imply a commitment by the enterprise to achieve what is stated.

The actions taken in support of these statements may provide evidence of diligence in working toward their achievement. Such actions need to be organized for consistency and effectiveness. Ad hoc actions may adversely affect evidence of diligence for the enterprise, whereas consistent actions help to deploy best practices and positively affect evidence of diligence. These enterprise-wide actions are more efficient because they are determined in advance of execution on each program and make efficient planning possible. Then the activities are tracked, and evidence can be produced on time for compliance with the enterprise safety process.

Governance

In order to ensure that such evidence is produced, the enterprise can emphasize the existence of a governance authority by assigning metrics to the process that are reviewed by that authority. Collection of these metrics can be controlled by the auditing and assessment group to ensure objectivity. As the assessments are performed on the work products of the safety process, the assessor can note whether they are completed on time when they pass assessment; the safety manager can let the assessor know if work products do not require the highest level of independent assessment. These data are then used to compute on-time metrics for each program, which are also consolidated by product line to be reviewed by the safety governance authority and enterprise executives.

Confidence in the completeness of surveillance employed by this governance authority is shown by ensuring that the instantiation of the system safety process for every product is included in the surveillance metrics. If one program in a product line is complying with the safety process while another is not performing any of the safety process, then the metrics will include this data for the noncompliant program at the earliest milestone. This enables corrective action to be directed for the noncompliant program, and the corrective action is reviewed until recovery is achieved.

But if, for example, safety-related products of one division are included in the system safety metrics while safety-related products of another division are not included, confidence in the completeness of surveillance will be reduced. If governance only has authority over the included product lines, completeness may not be achieved unless directed by enterprise executives responsible for all product lines. This may require escalation by the governance authority of the compliant product lines to the enterprise executives who have the authority to direct the other product lines to be included. Then consistency can be achieved and enhance the evidence of enterprise diligence.

Likewise, if one type of safety-related product is included in the surveillance metrics but another is not, then confidence in the completeness of the surveillance employed may be reduced. This may also require escalation to the enterprise executives who have the authority to direct that the product lines be included. Inclusion of metrics for all the safety-related products in the surveillance reviews provides evidence that is useful for the enterprise to achieve the stated safety goals.

Metrics

Achieving these goals requires control of the safety process across the enterprise. It is difficult or impossible to control what is not measured. Consistent review of metrics also supports a safety culture that ensures that the emphasis put on profit or schedule is not greater than the emphasis put on safety. The review of safety metrics by enterprise executives can maintain the required emphasis for this support of the safety culture.

The design of the metrics should ensure that the system safety requirements are complete and verified on time for each product. Milestones are chosen such that a significant milestone precedes the design and another significant milestone precedes the start of design verification. Methods for ensuring elicitation of requirements, such as systematic safety analyses, are required to be complete by the metrics at a milestone before verification. For any deviation, mitigation is ensured and recovery is tracked.

Such a program of executive surveillance of system safety performance using metrics can include measures to ensure objectivity. This is achieved by using the auditing and assessment department to collect the metrics data, as previously described in this chapter. The data are reviewed by the auditing and assessment department but compiled by quality assurance for executive review. Either method can ensure objectivity.

The metrics attribute may be binary or more complex. A more-complex metric can measure the amount of lateness to provide insight into programs or to help prioritize the deployment of resources for needed recovery. For example, a metric that measured lateness by domain as well as by program can help clarify the status-specific resources being applied to the safety process in particular product lines and across the enterprise. However, even a binary attribute requires expert judgment for determination. The completion of work products for system safety needs to be judged as adequate by an expert in system safety who has the required independence, such as another safety expert on the program or the independent safety assessor. This judgment should be independent of the product authority or program management, unlike other schedule achievements that are determined directly by program management. For metrics concerning safety, this avoids the appearance of a conflict of interest. This independence is also consistent with the requirements of ISO 26262 for assessment.

Still, after determination of the metric status, further reporting of this status is best done by those responsible for its achievement. This helps to establish ownership by making it transparent to the enterprise executives. Reporting of safety metrics by quality or the independent safety auditing and assessment leadership can cause conflict with the responsible executives, which clouds the true status. If the responsible executives, such as the engineering executives, report the status to the enterprise executives that has already been agreed on with the system safety assessors, meaningful review can be achieved. Emphasis is intended to be on achievement, with the metric serving as a tool.

Process Documentation

The procedure for using metrics as a tool to ensure diligence in the execution of the system safety process is included in the process documentation and may include operational definitions of the metrics to be collected. The method and authority for employing these definitions are clearly described; the period over which the metrics are collected, any required latency for the compilation of the metrics, and other data concerning distribution may be included if helpful or may be described separately to facilitate improvement or exceptions for special situations.

This process documentation requires approvals to be released. These approvals may include program stakeholders and product lines included in the metrics and representatives of the governance group. Representatives of enterprise quality leadership can also be included in the release to ensure that metrics reporting is efficiently and clearly understood when viewed in the context of other metrics that are reported for executive review. Using the enterprise process for release of this system safety process documentation, including metric definitions and determination, helps calibrate stakeholder expectations concerning the review criteria of executive management. This is important not only to avoid conflicts, but also to help the reporting product line executives be prepared and successful when they direct their product line compliance and report the results. Further training and briefing of these executives concerning process, metric definitions, and review criteria is necessary.

Audit reports and escalations provide timely updates as well as clear recommendations in order to help executives be successful with respect to safety process compliance. Further metrics with a more granular definition are collected to support the stakeholders achieving favorable summary metrics for executive review. In this way, they have a deeper understanding of the metrics with respect to their product lines than the metrics reported at the enterprise level.

Product line executives may have senior management and directors report to them concerning both sets of metrics in order to successfully direct process compliance in their product lines. These reports may be more frequent than review by enterprise executives and can, for example, support creating an effective recovery for the late execution of a safety-related analysis during the design phase. This recovery is initiated within the product group and is in process or completed before reporting to the enterprise executives. Early issue identification, program audits, and frequent reviews within product lines support the necessary focus, priority, and resource allocation to effectively execute a successful recovery.

Tiered Metric Reporting

Consider the more-granular metrics that measure on-time, approved system safety work products at intermediate milestones of product development, as shown in Figure 9.1. These milestones are at program phases that are identified in any automotive process for program management and include initiation, concept, design, verification, and validation. Each phase has different work products that are compiled progressively to provide evidence of compliance with the safety process and for inclusion in the safety case. In preparing for intermediate milestone reviews, the leader of the design function responsible for the review can prepare a recovery plan, if needed, to be included in that review, perhaps with the endorsement of the assessor from the central system safety organization. Responsibility is clear for the recovery plan and its execution. The assessor and auditor can support the success of the responsible leader by identifying what is needed and providing expert advice on how to get it done, including specialist resources that can help.

Early detection of compliance issues provides the maximum time for resolution and recovery. Rapid deployment of expert resources increases the likelihood that timely resolution will ensure compliance with the safety process. If recovery is completed before the summary metrics are prepared, the summary metrics will present an improved status and support a favorable review as a result of the prompt and effective action of the product group.

If recovery is later, then progress is reported at the executive review. The exact issue, the actions being taken, the resources being deployed, and any hurdles the program personnel must overcome can be summarized concisely, along with progress being made and a projection of the potential recovery timing, which may require executives to prioritize resources. Information is available about what is needed and what results are expected based on current or requested resource prioritization. An informed decision can be made about whether additional resources are warranted.

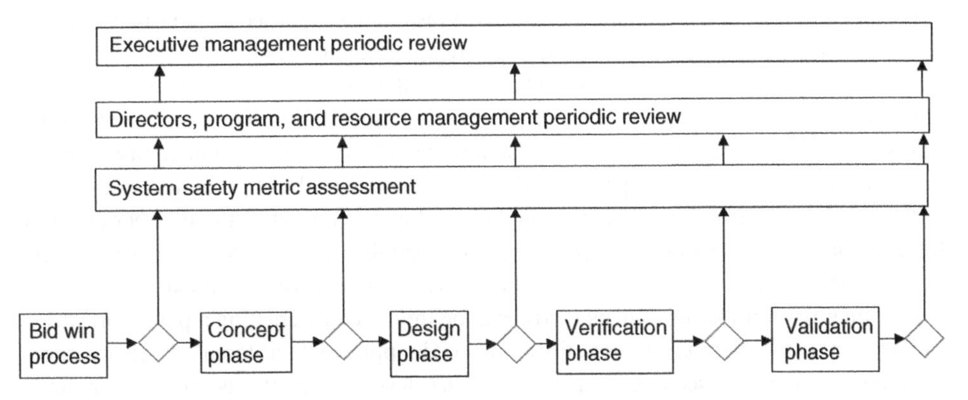

Figure 9.1 Example of system safety metric structure.

Either way, having more granular metrics for use by resource managers and directors supports improved control of safety processes. The additional information enables finer resource allocation and earlier execution of needed recoveries. Prioritization decisions are better informed and more successful because they are data based. This supports execution of the safety process on all safety-related product launch programs, because the launch programs have the same milestones and metrics. Each can execute the process efficiently using feedback from auditors, assessors, metric preparation, and the metrics themselves. Reviews are executed more frequently than the milestones, which allows management of work product completion and recovery (if needed) in time to have the most favorable review possible. Thorough documentation of the process and operating definitions for the metrics helps avoid nonproductive metric discussions. The metrics presented are understood, discussions focus on achievements and needed actions, and the emphasis is on execution.

Such metrics, structured as illustrated in Figure 9.1, improves the confidence of potential customers in the safety process. Customers can see the systematic method being used to ensure that the safety process is executed. Executing this process supports achieving system safety for the product to be delivered to the customer. The safety culture is demonstrated by the enterprise's overall commitment to achieving system safety; this also provides evidence confirming that the customer has taken safety into account when purchasing safety-related systems as required by ISO 256262.

Tracking progress to support management of improvement efforts can be demonstrated by plotting the results of periodic reviews. These results can include monthly metrics data plotted over time for overall enterprise performance as well as plots of the overall metric data for each product line. Further plots can include enterprise performance for each program milestone and the performance of each product group in a product line. More granular metrics increase transparency, especially for lateness and causes of missing any milestones, and improve customer trust.

Use of Metrics

The use of metrics to manage a process to achieve execution goals is well known in the automotive industry. Manufacturing metrics measure machine uptime as well as productivity per manufacturing line and shift. Further transparency is gained by measuring causes for downtime such as machine setup, machine repair, and quality checks.

Using metrics to improve product quality is also well documented, such as the number of return tickets issued and the number of no-trouble-found results. The number of return tickets measures how many times the decision was made to return a product based on the performance observed versus expected performance, as well as the effectiveness of the diagnostic and repair instructions. The no-trouble-found metric measures the number of times this decision was incorrect or the issue was not repeatable. Both of these metrics support managing the enterprise toward quality-related goals. If the decision to make a return was correct, then the number of returns metric can be improved by creating a pareto of causes leading to returns, such as component failures or inappropriate automated performance. Then the pareto can be used as guidance to prioritize development efforts to correct or eliminate these causes by improving the product in the field or the next product generation.

These metrics are reviewed periodically by multiple levels of management. Enterprise management seeks to understand the company's competitiveness with respect to quality sufficiently to either confirm the direction being taken or direct that resources be deployed to correct the direction.

More-frequent reviews are held within each product organization of the enterprise, to assure effective execution of the product quality process. These additional reviews collect metrics at product line manufacturing sites that contribute to the overall metrics for the enterprise. The aging and resolution of field issues are measured so that responsiveness of the product line to the customer is verified. In addition, metrics are reviewed to ensure the successful execution of the engineering quality process. Such management of engineering quality may support improving the level of maturity in the software process improvement capability and determination (SPICE) model [16], which is audited.

Likewise, metrics that measure the state of elicitation, verification, and validation of system safety requirements throughout the program, with multi-level management review, provide confidence that system safety diligence is achieved. There may be hundreds of thousands of system safety requirements that need to be elicited and verified on all of the safety-related programs in the enterprise. Missing any one of these could lead to a safety-related incident in the field and recall of the product.

Metrics provide a level of abstraction that supports effective management of a process to systematically ensure completeness of system safety requirements and compliance; this is supported by tracking the periodic metric results to demonstrate progress toward targets agreed on with executive management. These targets may be for the timeliness of executing the process that ensures completeness of system safety requirements and compliance. However, no individual program's safety case can be endorsed by the system safety assessor without evidence that the risk is not unreasonable. When these metrics show that progress reaches a competitive result, the metrics are used by automotive suppliers to search for new business for safety-related products.

Internal metrics for overall enterprise compliance with internal system safety process are considered confidential. No nation recommended including such metrics in the standard ISO 26262 for automotive functional safety or ISO PAS 21448 for SOTIF. No standardized metrics for automotive system safety are published, but a strong, successful commitment to system safety is an important discriminator to reduce potential safety risk to the VM. Every VM is aware of the potential loss of resource and reputation that may be incurred as the result of safety-related defects in products released for series production and public sale. Measures instituted by suppliers to reduce such risks are considered valuable, so VMs welcome transparency into the effectiveness of these measures, as shown by metrics.

Likewise, VMs can use improved metrics to show diligence to regulatory agencies or other forums where a demonstrated commitment to system safety is advantageous. These metrics, collected internally by the VM, can also be structured in a manner similar to Figure 9.1. This tiered metric reporting can demonstrate the same systematic diligence discussed earlier by executives of the VM enterprise to ensure the completeness of system safety requirements and compliance in order to protect the general public. Such diligence is viewed favorably by regulatory agencies and forums, and favorable metrics shared with VMs by suppliers further corroborate this demonstrated commitment. The VM may be credited with selecting suppliers that have demonstrated diligence and continuous improvement in the execution of an effective automotive safety process. The metrics are evidence of managerial commitment to surveillance in order to sustain this execution.

Such summary metrics are examined efficiently in the context of other business discussions: their value is considered in the allocation of resources and may also be studied by marketing and sales personnel. Additional explanations can be provided if requested. In this way, metrics can start useful discussions.

Consider a case where a VM is considering granting a business award for a new SOTIF-related feature, such as a level 4 automated system for urban use in selected geofenced environments. Such a system is also within the scope of ISO 26262 for automotive functional safety, so there is a standard requirement to consider the safety capability of potential suppliers. This consideration is initiated in the request for quotation (RfQ). An RfQ is issued that describes the intended application and some higher-level system safety considerations, which may include compliance with ISO 26262 and ISO PAS 21448. In addition, there is a request for support in SOTIF validation with analyses, simulation, track, and public road testing. Validation goals may be included, or a request can be made for support defining the validation goals.

In addition, a proposed Development Interface Agreement (DIA) is provided to organize and agree on the system safety responsibilities and shared documentation expectations prior to awarding the business. This DIA may require the supplier to perform a hazard analysis, including both functional safety and SOTIF considerations, by a date very soon after the contract award. There may also be requirements in the DIA for the supplier to share sensor limitations and to provide analyses showing how these limitations are accommodated safely early in the program design phase.

Traceability of requirements and review expectations are also provided in the RfQ. The VM may have a database or format for these requirements, or may require extensive audit, review, and approval of the supplier's requirements database. These reviews may include checking that the child requirements actually satisfy the parent requirements completely, that the verification methods are sufficient to verify each of the requirements adequately, and that the requirements and verification are complete.

Suppliers receiving the RfQ can map the documentation expectations to their standard process and list the equivalent documents to be provided to satisfy the VM's documentation requirements while not adding cost to the program execution. Suppliers may, for example, have several work products that, when considered together, satisfy all the requirements for a work product required by the VM. Alternatively, the RfQ documentation requirements may be based on ISO 26262, with some additional work products appropriate to ISO PAS 21448, while the supplier's process is based roughly on ASPICE base practices. These work products can be supplemented with system safety evidence and additional work products as required. In this case, a single supplier work product may satisfy multiple VM work products.

Traceability of system safety requirements is agreed on by all suppliers providing a quotation, as well as competitive automotive quality metrics. Any supplier engaging in development of level 4 automation can be expected to have a mature process for requirements capture and management. Competitive automotive quality metrics can also be expected from any suppliers receiving an RfQ for systems such as advanced safety-critical products.

Suppose that one supplier's bid package summary includes system safety metrics that have been reviewed periodically by executive management. Such data is supplemental and is not required by the RfQ. An explanation provides information concerning the operating definitions of the metrics as well as how these definitions are used in metrics determination to ensure objectivity in their application. The review process is also described, such as is shown in Figure 9.1, along with how the feedback is used. These metrics demonstrate a level of competitiveness and continuous improvement of on-time elicitation, verification, and validation of system safety requirements for all products across the supplier's enterprise for a significant period of time. Such experience in managing safety-related product development is valuable and demonstrates the leadership of this supplier in diligence in the system safety domain. Such leadership is appropriate and preferred for the development of a level 4 automated driving system. Execution of requirements to verify the sensors and algorithms as well as completing system verification in areas 1, 2, and 3 of ISO PAS 21448 may need to rely on such system safety leadership and experience. In addition, the system safety metrics demonstrate that the infrastructure and management capability are in place and available to support the program being bid. This is particularly important if the system is expected to have high visibility in the automotive industry and be the focus of regulatory scrutiny. The additional governance of system safety by this infrastructure may provide more favorable optics. Thus, including such metrics can support VM confidence and support business decisions.

10

Summary and Conclusions

Background

The previous chapters have discussed automotive system safety. Automotive system safety includes functional safety, safety of the intended function (SOTIF), and safety from other hazards that are not functional. The definition of *safety* that is commonly used by automotive safety specialists who reference ISO 26262 includes "no unreasonable risk." This definition is also referenced in ISO PAS 21448 for safety. It is appropriate to system safety since the unreasonable risk referenced is agnostic to the cause of this risk – the cause may be a functional failure or nonfunctional.

This definition links achieving safety to determining whether the residual risk is unreasonable: that is, whether the risk is greater than the risk that is acceptable based on societal norms. If a product puts the general public at a risk that is greater than those norms consider acceptable, then the product is not considered safe. It is assumed that the risk accepted by society today is consistent with societal norms; this includes risk that can be determined for a particular type of accident that is caused by a system's normal or malfunctioning behavior. It also includes the risk that a system presents to society if the methods used to develop that system – including elicitation of and compliance with system safety requirements – are not at least as comprehensive as those used for similar products in the field.

The risk imposed on the general public from exposure to the automotive products being used today determines the risk that is not unreasonable. This is the risk that the public must accept when operating or being exposed to automotive products. Members of society are aware that there is a risk of harm; this risk is mentioned regularly in the news and by other sources of information available to the general public. Automotive products are acquired and used, and that is evidence that the risk has been accepted and is not unreasonable.

System Safety Is More than Functional Safety

Automotive products that are being prepared for release are expected to be safe. This expectation is held by consumers, manufacturers, and regulators of automotive products. Therefore, new products must not impose a risk on the general public that is greater than the risk imposed by automotive products today. (Determining this risk is discussed in

Automotive System Safety: Critical Considerations for Engineering and Effective Management,
First Edition. Joseph D. Miller.

Chapter 1). Any product in the field that exhibits a greater-than-normal risk for products of its type is recalled, even if no injuries have been traced to the risk. If the risk is traced to a significant lack of diligence by the producer, the producer may suffer additional penalties. Thus there are compelling commercial reasons to satisfy the safety expectations of potential consumers of automotive products.

Safety Requirements

The determination of risk includes studying accident and field data, communicating product limitations, and considering ISO 26262 and ISO PAS 21448. Accident and field data are available from statistics collected by governments as well as from field data collected by vehicle manufacturers (VMs). Consideration of ISO 26262 and ISO PAS 21448 includes compliance with the requirements and recommendations of these standards. It also includes eliciting requirements from three sources: requirements that flow down from parent requirements; requirements obtained from systematic analyses of the system, hardware, and software; and requirements derived from the assumptions of others. Parent requirements include safety goals derived from a hazard and risk assessment, functional safety requirements from the functional safety concept, and specification of the top-level required safe behavior of the intended function. This flow is illustrated in Figure 1.1: it includes the flow of requirements to responsibilities in the development interface agreement (DIA), and the resulting communication of additional requirements during the execution of the DIA.

Failure to consider any of these requirements may impose an unreasonable risk on society. Perhaps a safety requirement that could and should have been considered was not. Failure to verify any of these requirements may impose an unreasonable risk on society: if requirements that are identified and known to be safety related are not checked, the product's compliance with these requirements is unknown. Following a systematic process can prevent missed or unverified requirements, and process enforcement ensures diligence.

The requirements for automotive safety significantly exceed the requirements for functional safety set out in ISO 26262. Automotive safety includes compliance with the requirements of ISO 26262: functional safety is necessary to prevent unreasonable risk to the general public but is not sufficient. The requirements of ISO 26262 are the result of years of collaboration by international experts and merit compliance. During these years there was considerable debate and consensus regarding the requirements for functional safety. After the first edition had been in use for several years, the standard was revised and released in an improved and more comprehensive second edition.

These requirements are necessary for the achievement of functional safety, and evidence of compliance indicates diligence with respect to functional safety. For a vehicle being launched, this includes the evidence compiled by tier 2 suppliers, tier 1 suppliers, and the VM.

Likewise, the requirements for automotive safety exceed the requirements for SOTIF set out in ISO PAS 21448. While SOTIF originally started as the work of a small group, the final requirements are the result of consensus among a large, international group of safety experts committed to providing guidance that is applied to various levels of automated driving. The requirements of ISO PAS 21448 are the result of years of collaboration by international experts and merit compliance; all domains involved in achieving safety for various levels of automated driving were represented in compiling this standard. These requirements are necessary for the achievement of SOTIF, and evidence of compliance indicates diligence with respect to SOTIF. Each product developed that falls within the scope of the levels of automation covered by ISO PAS 21448 (or higher) needs to show evidence that the requirements of ISO PAS 21448 were considered and that the applicable compliance has been achieved.

Critical consideration of the three sources of requirements for both functional safety and SOTIF are needed for automotive system safety. These again include SOTIF requirements that are flowed down, requirements that are derived from analyses, and requirements resulting from the assumptions of others. A systematic process supports ensuring that the requirements and verification are complete, thus supporting the achievement of system safety.

Safety Process

The system safety process can be divided into two processes: one that targets product safety, including SOTIF, and another for functional safety. Safety experts on the ISO 26262 committee who were from enterprises that divide system safety into these two processes favored putting the SOTIF in a separate standard from ISO 26262: they were concerned that the SOTIF requirements would not be accepted by the product safety process because ISO 26262 requirements were not addressed by the owners of the product safety process. The experts from the functional safety process that opposed including SOTIF in ISO 26262 were responsible for the functional safety process and were concerned that they would be expected to assume the additional responsibility for SOTIF that should be handled by the other group.

Different experts may be involved in each process. Functional safety experts familiar with ISO 26262 are the main experts for the functional safety process. Some experts may also be system, software, and hardware experts as well as specialists in safety analyses for these domains. Efficiency is gained by combining SOTIF and functional safety into one system safety process, as discussed in Chapter 3. On the ISO 26262 committee, the experts who supported including SOTIF in ISO 26262 were from enterprises that combined, or would combine, the safety process for SOTIF with the safety process for functional safety. It was more convenient for them to have the requirements in one standard: consistent definitions would be ensured, and there would be no conflicting requirements.

Many of the experts who developed ISO 26262 also developed ISO PAS 21448. These experts are familiar with similarities between the two processes and what needs to be added to functional safety work products and analyses to comply with SOTIF requirements. This streamlines the required work for compliance with both standards. Having a single process also can simplify achieving a single process for enterprise-wide product development. Training can be coordinated to address both standards in the process more efficiently; additional safety work required by the different engineering domains is minimized. Therefore, deployment enterprise-wide is less intrusive to the programs than two separate processes would be.

This approach also simplifies communication with cybersecurity personnel and resolution of requirements conflicts, as illustrated in Chapter 8, Figure 8.1. Such communication is required by ISO 26262 and is extended to SOTIF. When the processes are combined, communication between safety and security comprehensively includes functional safety and the intended functions, including cyber attacks targeting the vulnerabilities of both. The attack surfaces are nearly the same for SOTIF and functional safety. Potential conflicts are avoided.

Five Criteria for a Successful Safety Organization Are Key

An effective system safety organization supports successful execution of a system safety process. Every organizational choice is a compromise and has different advantages and disadvantages for implementing the system safety process effectively and sustainably. To ensure the success of the system safety organization, five critical considerations must be met:

1) The organization must have the talent to perform the safety tasks. This talent can be acquired as well as developed internally. It is needed to support the elicitation and verification of requirements as well as to perform safety analyses and communicate with safety managers of other organizations. Also, teaching system safety, mentoring, auditing, and assessment require system safety talent.
2) System safety must be integral to product engineering. This facilitates including system safety analyses in the concept phase and eliciting system safety requirements for design and verification.
3) There must be a career path for safety personnel to retain them in a safety role. There is a high demand for experienced, competent system safety personnel, especially in the automotive industry. When a high-potential, competent system safety expert gains experience in the enterprise, this expert also gains value outside the organization. Retention requires internal opportunities to be competitive: a career path is required.
4) The safety process must be owned by program management so that the tasks can be planned, resourced, and executed. For the safety tasks to enjoy the same priority as other critical tasks in the launch program, they need to be scheduled and made visible by the program manager. Then they can be monitored so that measures are taken to ensure execution. Issues are tracked and escalated if necessary.
5) There must be periodic executive reviews to ensure that the process is followed. Both product line reviews as well as enterprise-level reviews are needed, to empower the organization to follow the safety process. There may be conflicts for resources, and executive review ensures that these conflicts are resolved so the reviews are favorable.

System safety organization success requires all five – missing one leads to failure.

Chapter 2 discusses (and compares, in Table 2.1) the effectiveness of three organizational structures shown in Figures 2.2, 2.3, and 2.4. There may be other considerations to trade-off, when choosing the organizational structure, that are in conflict with a structure that is optimized for the critical considerations for system safety in the previous list. In general, alternative 3 scores highest due to supporting the acquisition, development, and retention of system safety talent better than the other two options. This alternative includes the enterprise's safety experts in one central organization; safety managers and safety experts are deployed to product lines and programs from there. This approach improves mentoring of direct reports and provides the most competitive career path for system safety personnel.

The other two organizational structures are more integrated with the product engineering organization but may experience less mentoring and more attrition. Safety managers and experts report to non-system safety personnel to whom they are assigned. To advance their careers, safety experts must either leave system safety, move to the central auditing and assessment group, or leave the enterprise. Labor efficiency and personnel development are also better supported by alternative 3 due to the potential for load leveling and job rotation. Peak demand is addressed by reallocation of resources by the central organization, which also broadens the scope of program experience for the safety experts. Care must be taken to mitigate disadvantages of organization 3. This is true of all organizational choices – there are always trade-offs.

Auditing and the Use of Metrics

Auditing

Chapters 2 and 4 examine several alternative approaches for auditing the system safety process and assessing the system safety work products within an enterprise that produces automotive products, either as a supplier or a VM. Chapter 2 discusses how the activities of the central organization are affected based on the organizational choice. There are management responsibilities concerning the safety managers and experts within the product lines for alternative 3; the central organization also deploys these experts.

Chapter 4 discusses the use of external auditors who are commissioned by the program, versus using an internal auditing organization. External auditors may be independent; but if the program being audited hires them, an internal organization not financed by the product group is better. The internal organization can supplement its workforce by hiring external auditors and still maintain financial independence. The audits and assessment can serve as teaching opportunities and support the adoption of best practices throughout the enterprise. This is enabled by having the auditors cover more than one product line: networking and examples can then be shared efficiently across product lines, especially to support recoveries.

Audits and assessments produce value by ensuring that the risk of harm to the general public of releasing the product is not unreasonable, and that evidence supports an argument of diligence. Any issues that may prevent this are identified early and escalated if needed. Concerns by the product team can be brought to the auditor's attention without risk of repercussions. The assessors and auditor are knowledgeable about the relevant standard. Compliance with accepted standards strengthens this argument of diligence. This compliance is noted in the safety case approved by the assessors. Audits and assessments support compliance.

Metrics

The central safety organization can also support the safety culture through metrics that measure compliance with the safety process and enable executive review. The transparency to executive review encourages behavior that achieves positive metrics for reporting at these reviews. An early milestone requiring the completion of a successfully assessed work product may be chosen for a measurement, to indicate that the safety process is instantiated successfully on the program and that early progress is satisfactory. Failure to meet this milestone can adversely affect the desired evidence of diligence. In addition, a milestone should be chosen to measure whether all the analyses necessary to elicit derived safety requirements have been completed in time to be verified, and still another to determine whether all requirements have been validated. The requirement to complete analyses to elicit requirements before verification is the most critical for engineering development. These domain-specific analyses, such as the hardware single-point fault metric or software hazard and operability study (HAZOP) software analysis, are used to elicit requirements for the design – if they are not completed by the end of design, then compliance cannot be ensured.

The validation milestone is intended to indicate that the safety case is completed. Other milestones are chosen to measure intermediate progress, as discussed in Chapter 4 with reference to safety audits. More granular milestones support reviews within the product line that are more frequent than enterprise reviews. Such product line reviews help ensure successful enterprise reviews, because timely recoveries are executed. Audits are thus supported by the metrics and help program management by providing actionable information about how to improve performance, if needed. Then program management can take appropriate steps. Senior management reinforcement helps performance improvement.

Vocal, active support of the system safety process, the evidence it produces, and system safety metrics by the legal department is critical to promoting a safety culture that embraces transparency. The legal department may enjoy an especially high level of respect and credibility within the enterprise, especially within engineering, concerning the importance and use of the safety case it supports. In this instance, the legal department is considered the customer for the safety case. Legal training for development and program personnel that includes liability topics related to safety is very helpful. Personnel working on the development of safety-related automotive systems, particularly innovative advanced automated systems, may have concerns with respect to their legal obligations as well as those of the enterprise. Having legal training provided by internal counsel adds credibility and encourages questions: attendees realize the stake that the legal department has in the enterprise with respect to the counsel provided, which is less apparent using outside counsel. As a result, understanding is deeper and internalized.

System safety metrics presented by program management to product line executives, as well as by product line directives to senior executives (as illustrated in Chapter 9, Figure 9.1), provide an opportunity for management to be successful by executing the system safety process. The expectation of credit and favorable compensation consideration cascades from senior executives to product line leadership, to program management, to program personnel for successfully executing the safety process while achieving other program milestones. Program management may receive critical support needed for this achievement. All share this common objective. Favorable metrics support hunting new business, and this success reinforces the value of meeting metric targets. The enterprise benefits.

Future Considerations for SOTIF

Chapter 3 discusses requirements of ISO PAS 21448. This specification represents a consensus reached over many years by safety experts from many nations representing VMs, tier 1 suppliers, tier 2 suppliers, and consultancies involved in the development of different levels of automated driving systems. While not specifically targeting the higher levels of automation, the requirements of this PAS are intended to be considered in the development and release of such SOTIF products. This extended use of ISO PAS 21448 was on the minds of the experts during its development and is so stated in the PAS.

The PAS is intended to be a predecessor to a new standard, ISO 21448, that specifically targets all the levels of automotive product automation. Work products that show compliance with the future ISO 21448 are included, which supports common interpretation as has been enjoyed with ISO 26262. Many of the same experts have been involved in both standards. This new standard is expected to include content from ISO PAS 21448 with improvements based on experience gained from its use. Many of these improvements will result in greater clarity and improved organization of the material, including the annexes and examples.

The three regions included in the verification and validation of SOTIF requirements from ISO PAS 21448 will still be applicable in standard ISO 21448, which is being developed. The clarity of the flow of verification activities is improved, and clarity is also improved by addressing context and feedback from a launched product. This may include the resolution of field issues. The principles of not incurring any unreasonable risk will still be applicable for SOTIF. Determining a validation target based on field data to ensure no unreasonable risk is addressed with considerable guidance and examples of such determination. This can significantly increase the effort for the validation of area three in ISO PAS 21448 for higher levels of automation. The use of simulation in addition to actual driving in achieving this validation target will receive considerable guidance in the future ISO 21448 standard, along with examples.

Searching for unknown triggering events becomes a more extensive effort as automation is expanded. Severity of potential harm is a factor in selecting validation targets, and the statistical confidence required to demonstrate that the target has been achieved is often higher to achieve wider acceptance by the general public. This higher statistical confidence is inferred from the target selected. Also, the amount of validation experience with other products by the product developer becomes more extensive. Targets met by previous similar systems are taken into account when determining achievement of the validation target. Guidance on this validation is expected in ISO 21448, and clarification may be provided.

The standard's annexes are being improved to use statistical methods to combine targeted verification with the extensive validation required to reduce the overall effort, which can grow massively as the scope of automation increases. This drives innovation to achieve the required confidence pragmatically. For example, by understanding sensor limitations, scenarios can be selected to both train machine recognition as well as verify the validity of this recognition in similar but different scenarios. These scenarios are in the same equivalence class bounded by the same edge and corner cases: if validation effort targets scenarios that test the limits of recognition, it is inferred that the other scenarios in the equivalence class have been validated. These actions improve the confidence in this validation as compared to a random validation effort. Worst-case scenarios are validated with a sample of other scenarios for added confidence. Credit is taken by the product development group for this confidence, and validation becomes more efficient.

Machine Learning

Machine learning is becoming more prevalent in products with higher levels of automation. This provides classification that is useful in avoiding false triggering while tuning appropriate system responses. The new standard ISO 21448 will provide more guidance on the safety of machine learning, to avoid systematic faults that machine learning may possibly induce. A safe-fail operational state should be ensured. The driver should not be expected to take over immediately in case of a failure – reaching a minimum risk condition is required. This requires that the system safety mechanisms detect such a failure with high confidence.

Several alternatives are possible for overall reasonableness, detection of failure modes, or a combination, and a strategy is required to execute the alternative chosen. Higher levels of automation may also require over the air (OTA) updates that require careful attention to security as well as system safety. Executing OTA updates safely while ensuring availability is a consideration: some restrictions affect availability of the system for normal use. Driver or passenger behavior is a consideration, and a safe design that takes it into account requires validation. The design must avoid hazards that could be induced by a poorly timed or executed OTA update. Efforts beyond the requirements of the new standard may also be needed. Nevertheless, compliance with the new standard supports diligence.

Appendix A

IEC 51508 Compared to Typical Automotive Practices

IEC 61508 is a generic safety standard intended to be the basis for application-specific standards like ISO 26262 for the automotive sector. It sets out requirements and considerations that can be tailored for a specific sector standard. The objective is that all the considerations discussed in IEC 61508 can be considered by each sector and may aid in compiling comprehensive sector standards. Where a sector standard does not exist, IEC 61508 can be used directly; some automotive applications employed IEC 61508 directly prior to the release of ISO 26262.

The statements and judgements made in this appendix represent the opinion of the author based on his experience. They do not represent the results of a survey of the industry and are not intended to be absolute. Nevertheless, for personnel that may have had experience applying IEC 61508 or its derivatives in other industries, this review may be useful. IEC 61508 addresses functional safety only; it does not address safety of the intended function (SOTIF).

IEC 61508-1 – General Requirements

The first part of IEC 61508 addresses general requirements of the entire standard. There is no equivalent part in in ISO 26262, but there is a short section duplicated in each part of ISO 26262 that is only slightly tailored from part to part. The introduction of IEC 61508-1 does highlight some similarities to and differences from practices based on ISO 26262 in automotive. These are summarized in Chapter 3, Table 3.1, but there are further details not included in this table. For example, both ISO 26262 and IEC 61508 address methods for addressing the functional safety of electrical hardware and software, and indirectly provide a framework for addressing the safety of other technologies, such as mechanical parts and systems. Neither addresses safety concerns such as electrical shock. IEC 61508 and ISO 26262 have different criteria for assessing the risk of continuously operating systems like electric steering, and systems operating on demand like airbag occupant safety systems. However, in IEC 61508, a risk probability target is set, while in ISO 26262, the exposure criteria are adjusted based on frequency for on-demand systems. Both standards list methods for controlling faults, and both address avoidance of systematic errors. ISO 26262 is based on IEC 61508 as intended; but there are differences.

Automotive System Safety: Critical Considerations for Engineering and Effective Management,
First Edition. Joseph D. Miller.

A discussion of using the standard as a basis for other standards is totally irrelevant in the automotive sector, though in the second edition of ISO 26262, part 12 uses the rest of the standard as a basis for its application to the motorcycle sector. Likewise, the truck and bus sectors use ISO 26262 for their sector standard, and some tables in risk determination, definitions, and discussion are especially applicable to truck and bus. So, while the utility of IEC 61508 is partially enhanced by its generalness, ISO 26262 is quite specific to the automotive sector but includes some targeted expansion to other sectors. Thus, using IEC 61508 directly for the truck, bus, and motorcycle sectors has now been preempted by the second edition of ISO 26262. Each sector provided experts for this expansion, and ISO 26262 now serves these sectors.

In the automotive sector, communication and liaison with cyber-physical security experts is mandated by ISO 26262. IEC 61508 does not contain such a requirement. However, IEC 61508 is more specific: if a hazard can be induced by a cyber attack, then a threat analysis is recommended. If security threats are uncovered, then a vulnerability analysis is recommended. Further details are not included except by reference. Automotive has the expectation that these security analyses will be performed by security domain experts. An automotive security standard may be available; vehicle manufacturers (VMs) provide separate requirements.

Documentation requirements for functional safety prescribed in IEC 61508 are essentially fulfilled by automotive practices. The requirements of IEC 61508 are more general, allowing judgment about what is needed as long as the prerequisite information is available in time for each phase. Automotive practices provide normative work products, and prerequisite informant is mandated. Likewise, management of functional safety is very similar: responsible persons are assigned, configurations are managed, and recommendation from safety analyses must be addressed. Training and competence of safety personnel must also be addressed. This is true of automotive practices, as well. IEC 61508 specifically references periodic audits; it is not unusual for automotive suppliers and VMs to have periodic safety audits. ISO 26262 does not require periodic audits – it only requires audits.

IEC 61508 requires that maintenance and operations are addressed to not diminish functional safety. ISO 26262 goes further and requires feedback from operations to development if an issue occurs with manufacturing the product and retaining safety. Automotive safety practices require a quality organization, as does IEC 61508.

IEC 61508 requires a lifecycle, as does automotive system safety, and this lifecycle is the backbone of the safety process. However, the IEC 61508 lifecycle is substantially different from the lifecycle mandated by ISO 26262, which is prevalent in automotive safety. The fundamental underlying difference is that IEC 61508 envisions a lifecycle where the equipment under control is designed and built, then the safety systems are added, and then the system is commissioned. This is entirely different from the typical automotive lifecycle where safety mechanisms are included in the design and then the system is released for mass production. The concept of mass production is not apparent in IEC 61508, though it has been used in automotive, with some tailoring, for development of mass-production automotive products.

The lifecycle in IEC 61508 is very granular: for example, there is a separate concept phase, scope phase, and hazard and risk analysis phase. In the automotive lifecycle, these phases are included in the concept phase. Nevertheless, the requirements included in the lifecycle of IEC 61508 are in general fulfilled by the automotive lifecycle. Potential sources of harm are identified as well as the scope of the system boundaries in order to perform the hazard and risk analysis. However, when the hazard and risk analysis is performed in IEC 61508, risk is determined in a quantitative manner. ISO 26262 does not have a quantitative target for risk; ISO 26262 identifies exposure, and the likelihood of harm given an accident may be determined from traffic statistics. Also, in ISO PAS 21448, the risk in the field can be taken into account in order to determine validation targets. So, the automotive sector may quantify risk. Quantification is not like IEC 61508.

IEC 61508 has a separate requirements phase. There is not a separate requirements phase in the automotive industry: safety requirements are a fundamental consideration for creating and verifying an automotive design. It is the culture of automotive engineering quality to check that the requirements have been elicited and compliance verified. As was discussed in Chapter 4, auditing requirements traceability is a significant action in system safety audits. It is done continuously during development and is not a separate phase.

Likewise, the overall safety requirements allocation of IEC 61508 is not a separate phase in automotive: this allocation is normally accomplished during the design. A system engineering function is to determine which safety requirements are allocated to hardware and which are allocated to software. Then each of these domains derives requirements to implement the safety requirements assigned by system engineering. This is part of the design process, not a phase.

IEC 61508 has an overall operation and maintenance planning phase that is different from automotive common practice. This phase is be based on equipment under control, such as a manufacturing or power-generation facility. A comparable phase in automotive considers maintaining designed-in safety during manufacturing and maintenance. Very little of the guidance for this phase in IEC 61508 appears directly applicable to the automotive lifecycle. Operations and maintenance safety are considered, but the automotive context is different.

Validation is a phase in IEC 61508. Every automotive program has a task for validation, and this task must be planned. Validation is not normally a phase in the automotive system safety lifecycle; it is included in overall product development at the system level, and validation testing is performed at the vehicle level. Some of the considerations in IEC 61508 are relevant, such as planning and determining the pass-fail criteria, but most need to be tailored for automotive because the context is different.

As discussed earlier, IEC 61508 has a context of safety-related systems being installed and commissioned. Planning this installation and commissioning is included as a phase. The requirements are brief, and most are not applicable to automotive. Vehicle integration by the VM may be the closest to this type of planning. There is not a separate phase for this planning in the safety lifecycle, and there may not be separate safety systems commissioned in automotive.

Other phases included in IEC 61508 are requirements, realization, risk reduction, overall installation and commissioning, safety validation, operation, maintenance and repair, modification and retrofit, and decommissioning and disposal. None of these are phases in automotive. There are some similar tasks associated with the automotive lifecycle for maintenance and repair, but the others are tailored by the automotive standard for the safety of series production vehicles and systems. Safety in disposal is recognized as an automotive design and maintenance consideration, and requirements are elicited to ensure safety; it is provisioned in the design of the automotive product being designed. Automotive manuals contain disposal instructions, which is different from the IEC 61508 context.

The verification requirements for IEC 61508 very closely resemble practices in the automotive sector. The verification plan is expected to be developed in each phase, though the phases are different in automotive practices from the phases of IEC 61508. The verification plan for both IEC 61508 and automotive is to contain specific methods to verify each phase of development. In automotive, each safety requirement of each phase must be verified by a combination of methods appropriate for the requirement. This is indexed by the automotive safety integrity level (ASIL) attribute of the requirement in ISO 26262. Such selection of methods and techniques in each sector is anticipated by a note in IEC 61508. ISO PAS 21448 provides guidance on what shall be verified but does not index the methods. Rather, it defines three regions of verification. This is consistent with IEC 61508 verification requirements, so automotive verification satisfies IEC 61508.

The functional safety assessment as specified in IEC 61508 is very similar to the functional safety assessment commonly executed in the automotive sector. Both are required to issue an assessment report with a recommendation for acceptance, conditional acceptance, or rejection. In the case of automotive, the functional safety assessment is to be completed before the release for production and should be compiled incrementally throughout the automotive lifecycle. ISO PAS 21448 does not mandate a SOTIF assessment report but does have criteria for the same three outcomes prior to release of the functional safety assessment. In practice, this is combined into one process and report, or is completed separately in the automotive sector. IEC 61508 is prescriptive in specifying the functional safety assessment, the roll-up of incremental functional safety assessments, and the criteria for independence. All are similar to automotive practices; no significant conflicts exist.

While there are similarities in the documentation structure discussed in IEC 61508, there are also significant differences. In automotive, the functional safety standard ISO 26262 has specific work products that are normative for each phase. In contrast, IEC 61508 has no normative work products, but rather has an example of a documentation structure. Both automotive practices and IEC 61508 support documentation as not necessarily consisting of physical paper documents but rather being virtual evidence that is retrievable, understandable by humans, and archivable. IEC 61508 requires a document number, while ISO 26262 does not. This is significant and was discussed during the development of the original release of ISO 26262. The requirement for a document number was abandoned so that configuration management tools that are compliant with the requirements for configuration management, that only relied on a unique path to the virtual document, and that supported revision control could be used without the need for a unique number. IEC 61508 also requires configuration management for documents and includes examples of "binder" structuring depending on the complexity of the product. Common automotive practice is to maintain a documentation database that employs a folder structure in which different types of documents are cataloged. Common automotive practice is for the functional safety assessment to have links to these supporting documents without the need to first find the binder. The structure of the functional safety assessment supports the location of evidence for each phase of development. Reuse of documents is supported by reuse of the links. All the evidence of functional safety is retrievable in the automotive functional safety assessment: it is sometimes called the *safety case*," and it includes arguments and evidence.

IEC 61508-2 – Requirements for Electrical/Electronic/ Programmable Electronic Safety-Related Systems

IEC 61508-2 details requirements of the realization phase for electronic safety-related systems within the scope of the standard. It is intended to be used in the context of IEC 61508-1 that provides overall requirements for safety-related electronic systems and inherits requirements from IEC 61508-1. Also, IEC 61508-2 is intended to exist collaboratively with IEC 61508-3, which concerns the realization phase of safety-related software. In this sense, the context of IEC 61508-2 is representative of the relationships created in the development of an automotive product. The design of electronic safety-related automotive products exists in the context of an overall process that elicits system requirements including safety-related design requirements. During the design of an automotive safety-related product, further requirements are elicited by the analyses of the software design that concern the system, and requirements are elicited by the system design and hardware design that concern software. IEC 61508-2 establishes generic realization relationships, which relate to automotive products.

As in ISO 26262, the documentation and management requirements are detailed elsewhere in the standard. These were discussed in IEC 61508-1. Management of functional safety is expected to be continuous throughout development, as may also be expected in development of automotive safety-related products, and includes appointing a safety manager, and the use of audits and assessments of work products. IEC 61508-2 also requires output documentation of granular activities during each phase of the safety lifecycle but does not specify them as normative work products. IEC 61508-2 also considers the integration of an application-specific integrated circuit (ASIC) during the realization phase of the safety-related system. The inheritance of the safety requirements by the ASIC from the system and its symbiotic inclusion in the architecture is illustrated. A systematic design and verification process for the ASIC is included and referred to as a *V-model*. Such V-model development is common practice in automotive product development, including the use of ASICs. Revisions of each preceding task in the design side of the V-model due to subsequent task feedback as well as verification feedback are illustrated in IEC 61508-2. This understanding is common in automotive development; the V-model is generic.

The specification of design requirements in IEC 61508-2 is in terms of the subsystems, which is consistent with automotive practices for design requirements. It is distinct from automotive practices for functional requirements, which specify overall system functions without referring to design elements. These functional requirements for safety in automotive practices correspond to the required system safety requirements input for design requirements in IEC 61508-2. This allows an architecture to be specified that combines these design elements and requirements. Also, this specification is a prerequisite for the next step in IEC 61508-2: validation planning. Automotive practice is to also perform validation planning based on the requirements specification, though it may not be the second step in the process due to schedule and resource constraints concerning availability of concept samples for the program. Planning the concept phase testing may include hazard testing, but not formal validation of all system safety requirements of the design. These may not yet be implemented in the concept samples. Validation planning for automotive may vary in timing, though not in intent. Specifying validation methods for safety requirements is an automotive practice: all safety requirements are verified, and this verification may be audited.

When the design requirements become available, development starts and includes the development of an ASIC, according to IEC 61508-2. Such an ASIC must satisfy the requirements for the portion of the architecture that the ASIC is responsible for fulfilling, including hardware and software interfaces as well as the implementation of safety mechanisms. This is very similar to automotive practice except that the safety-related system is not totally separate from the automotive system. Safety-related and non-safety-related functions may exist in the automotive design as well as the ASIC. The next step in IEC 61508-2 is integration of the safety-related system. Automotive practice is to integrate the entire system. Both IEC 615052 and automotive practice conduct integration tests, and the results are recorded.

Then IEC 61508-2 moves on to installation, commissioning, and maintenance procedures. This differs from automotive practice because the automotive product, such as electric steering or a vehicle, does not normally have a safety system installed. The safety system is part of the automotive system and is included from the start. Any installation procedures are for the entire system being installed on the vehicle, like electric steering, not a separate safety system. This difference in the lifecycle was one of the motivations that drove the creation of ISO 26262 by all the nations' experts. The safety mechanisms are independent, but this independence is included in the architecture of the product.

Maintenance procedures are needed by both IEC 61508-2 and automotive practices to ensure that the safety of the system is preserved when maintenance is performed. This includes instructions concerning the safety of maintenance personnel. Validation of the safety system is the next step in IEC 61508-2. Automotive practice is to validate the entire system, including the requirements for safety; this system validation is performed at the vehicle level and includes not only functional safety requirements of ISO 26262, but also region 2 and region 3 requirements of ISO PAS 21448. The latter is not foreseen by IEC 61508-2; SOTIF is not in scope.

Modifications are covered by IEC 61508-2 for the purpose of retaining the integrity of the safety-related system. In automotive practice, any modifications of the system that may have safety-related consequences are in scope. Verification is always required for each phase in IEC 61508-2. Automotive practice is to verify the requirements as planned, not necessarily at the end of each phase. Both IEC 61508-2 and automotive practice require a safety assessment that is performed incrementally. In automotive, it precedes production.

The normative considerations for design requirements in IEC 61508-2 are nearly identical to automotive practices. In addition to the normative considerations for requirements in IEC 61508-2, automotive practices include bidirectional traceability. Given a verification method, all the safety-related requirements verified by the method are found. Automotive practices require that requirements have a unique identifier, a status attribute, and, for functional safety requirements, an ASIL. Also, automotive has more qualitative requirements, like being atomic. While automotive requirements are expected to completely implement the parent requirement, IEC 61508-2 has more details about the content of the requirements, such as the operating modes to be covered, proof-testing requirements, and requirements for the quality system. Covering all the operating modes is expected in automotive practice as well and is somewhat covered by J2980 for hazard identification. Then safety goals flow from these hazards. These are parent safety requirements. Proof testing is not usually expected by automotive practices, but the system is expected to have coverage of latent faults. Quality control is not expected in automotive; it must be built in.

Validation planning in IEC 61508-2 is consistent with automotive practices for validation planning for functional safety. It considers requirements, environment, and test facilities. It does not take into account SOTIF or the fleet validation testing that is traditional for automotive VMs, all of which are necessary to ensure the safety of the general public. These are not required for plant installations, and IEC 61508 does not consider them.

The system design and development considerations of IEC 61508-2 appear substantially consistent with automotive practices. Architectural constraints for safety are agreed on first. These architectural constraints are extended to any ASIC being developed, including redundancy or other equivalent measures. As it is for automotive design, when safety-related and non-safety-related functions are mixed, the integrity requirements for safety-related functions must be satisfied by non-safety-related functions unless enough independence is shown. IEC 61508-2 requires design analysis to elicit requirements, as was practiced by automotive even before IEC 61508 was issued. Derating is also prescribed by IEC 61508-2; this is automotive practice when necessary.

Systematic failure capability is metricized in IEC 61508-2. This is not common practice in automotive; however, dependent failure analysis (DFA) is common practice, and automotive engineering quality is commonly assessed using prescribed standards to determine capability. When necessary, diversity is applied in automotive practice and can mitigate systematic failures.

IEC 61508-2 introduces the concept of hardware fault tolerance to determine the necessary diagnostic coverage for each safety integrity level (SIL). This concept is different from automotive practices but not completely dissimilar. For functional safety, ISO 26262 implicitly assumes a fault tolerance of 1. This is implied by the single-point failure metric and the latent fault metric. The single-point metric requires that a safe state be achieved by a safety mechanism when a fault occurs. If a fault occurs in the safety mechanism first, then this is detected, and a safe state is reached. Thus two faults are required in order to reach a potentially unsafe state. However, automotive practice does not typically refer to hardware fault tolerance. For automotive functional safety requirements of ASIL D, the single-point failure metric and latent fault metric requirements are similar to IEC 61508-2 SIL 4 requirements with a hardware fault tolerance of 1. The safe-failure fraction of IEC 61508-2 is largely comparable to the single-point failure metric in automotive functional safety as long as all the parts considered are involved in achieving the automotive safety goal. IEC 61508-2 also provides guidance for safety mechanism timing, like the automotive fault tolerance time interval for continuous-demand systems. The IEC 61508 guidance for safety mechanism low-demand systems is largely fulfilled by automotive systems, such as airbag diagnostic modules. Automotive guidance is less specific, but actual practice is similar.

The guidance of IEC 61508-2 for architectural methods to achieve a SIL for the system is largely the same as the automotive practices for decomposition of requirements with respect to ASIL. Automotive practices consider an ASIL an attribute of a requirement, so combining redundant requirements is similar to combining the architectural elements to which these safety requirements are assigned. The actual SIL "algebra" is equivalent.

The requirements in IEC 61508-2 for quantifying the effects of potential random hardware failures are similar to automotive functional safety practices for managing the effects of random hardware failures, with some differences. IEC 61508-2 requires that the effects of hardware failures be managed for each safety function, similar to the automotive practice of managing the effects of hardware failures on each safety goal. Common practice in automotive safety is to combine some or all of the safety goals when computing the hardware metrics, and meeting the most stringent of the requirements. This is because, unlike the systems foreseen in IEC 61508-2, automotive systems are mass-produced, and suppliers often provide similar systems to multiple VMs. It is not unusual for VMs to define safety goals somewhat differently for the same system. By combining safety goals, suppliers can avoid the cost of redundant analyses that add no additional value. Only hardware changes, if any, need to be updated in the analyses. Also, IEC 61508-2 considers the effect of random human errors when a person is required to take action, and includes this probability in determining the required failure rate. This is not common automotive practice. Common automotive practice is to determine whether the controllability is adequate through vehicle testing. In some cases, other actions are taken to prevent unsafe operation if a safety-critical repair has not been performed. Countermeasures to human error are considered, but quantification is not common practice.

Both IEC 6508-2 and common automotive practices include measures to control systematic faults. However, the foreseen context of IEC 61508 is again evident. For example, testing for IEC 61508-2 requires that a decision be made about what can be tested at the supplier's site and what must be tested at the user's site. This is normally not the case in automotive practice except for the use of test tracks. Automotive suppliers normally complete system testing; validation is always on the vehicle.

IEC 61508 relies on the practices at IC suppliers to control systematic errors in common parts, such as microprocessors. ASICs may require special measures. This is somewhat like automotive, but automotive practice tends to rely on a safety manual for IC components used in a safety-critical application to formalize the documentation for both random and systematic measures. IEC 61508-2 references requirements in IEC 61508-3 for control of systematic faults in software. This is similar to the automotive practice of referencing ISO 26262-6 for such requirements. Automotive practice is also to employ engineering quality assurance (EQA) to support software quality.

Maintenance and operator requirements are substantially different between IEC 61508-2 and automotive practice. IEC 61508-2 primarily targets operators and maintenance personnel in a plant environment, though there is mention of mass-produced systems for the general public. Automotive practice targets drivers and shop personnel. Still, the principle of being robust against human error applies to both. Automotive practice relies on procedures and design – redundant independent operator checks are not relevant to drivers. Maintenance checks may rely on maintenance diagnostics, and IEC 61508-2 is consistent with this.

IEC 61508-2 requires that a safe state, or actions to achieve a safe state, be specified in case a dangerous fault is detected. This action may include repairs within the minimum repair time. The system must then be implemented to meet these requirements under all operating conditions. Continuous-use and low-demand systems are included, which is consistent with automotive practices. Also, IEC 61508-2 requires that any component claiming to meet IEC 61508 is required to have a safety manual: it must provide evidence of compliance, what is needed to use the component compliantly, and sufficient data to allow the component to be included in the SFF of the system. This is like automotive practices, though the safety manual can be provided even if it is not always required. Still, automotive practice is to construct a safety case that will consider compliant components. The detailed requirements of IEC 61508-2 are substantially fulfilled by automotive practices. References are normally provided to retrieve evidence. This supports a safety argument.

Both IEC 61508-2 and automotive practices allow credit to be taken for a component or system that achieved safety in actual use. In IEC 61508-2, such a system must have demonstrated a probability of failure low enough for the SIL claimed. However, in automotive practice, the candidate must have demonstrated a risk an order of magnitude better than is needed for a comparable SIL in IEC 61508-2. Otherwise, such a system could potentially have experienced a safety recall and still be claimed proven in use (PIU). The other requirements for the candidate, such as specification, change management, and suitability for the current application, are essentially the same as automotive practices. PIU is almost never used by VMs and is seldom used by automotive suppliers for systems with higher ASIL requirements, because of the large volume needed to demonstrate compliance. Components meeting lower ASILs may satisfy PIU, and this has practical value.

IEC 61508-2 discusses additional requirements for data channels. The types of failures and reliability requirements are consistent with automotive practices. The *white channel* (compliant channel) and *black channel* (compliant interfaces) of IEC 61508-2 are not typical descriptions used in automotive practice. Automotive often uses *end-to-end* protection to transmit data through a runtime environment that is not secure. Also, communication channels offering a more secure architecture for communications are also used in automotive practice. IEC 61508-2 also discusses requirements for system integration, including required documentation for the system configuration, test configuration, and results. There is a significant list of requirements, including the impact of changes; the use of equivalence classes to reduce the number of test cases is allowed. These requirements are consistent with automotive practices. Automotive practices also consider system calibrations, which cover various vehicle configurations.

IEC 6158-2 also has requirements for operations and maintenance. As is true in automotive practices, these requirements are to maintain the safety that has been designed into the safety system while being manufactured and during maintenance (either due to failure or due to periodic preventative maintenance). Both are applicable to automotive; the considerations for documentation, tools, and analyses are similar. Automotive practice is to collect data for possible PIU consideration. This is not specifically mentioned in IEC 61508, but the practices are similar.

IEC 61508-2 has requirements for validation in order to determine compliance with the functional safety requirements specified. This is common automotive practice. IEC 61508-2 requires that every possible combination of plausible event be validated unless independence can be shown. There are significant details specified concerning considerations and documentation; issues are managed, and suppliers share validation data with customers. Validation occurs before launch in automotive; everything else is consistent practice. IEC

61508-2 also requires changes to be managed to maintain safety. The documentation is updated, and the required revalidation testing is performed. This is like automotive practices.

IEC 61508-2 has verification requirements that are to be applied to each phase in order to ensure that the requirements for that phase have been fulfilled. This includes verifying that the system design requirements satisfy all the safety requirements. Automotive practice may not include verification after each phase of development; however, traceability of requirements is a continuous process for system safety in automotive, including checking that the child requirements satisfy the parent requirements, including safety. Requirement conflicts are resolved, such as safety and security requirements. IEC 61508-2 discusses testing after each phase, but automotive practice is to have testing appropriate for each phase and complete verification after the design phase. IEC 61508-2 documentation practices are like automotive; fault detection is verified.

IEC 61508-2 contains an annex of measures to control random and systematic faults with various levels of coverage and recommendation. Automotive practice is to consider similar tables in ISO 26262. The latter standard lists low, medium, and high coverage, like the 60, 90, and 99% in IEC 61508-2. Similar claims are made for diagnostic coverage in automotive practice; the explanations were updated in ISO 26262 for automotive applications. The basic claims of coverage are used, and improvement in coverage is achieved by multiple measures. The methods for each of the tables are similar to automotive. Use of the tables is common practice, and automotive practice is consistent.

Likewise, IEC 61508-2 has tables that provide guidance for managing systematic faults in all phases of the safety lifecycle, including operation, maintenance, and the training of operators. Measures of this type target installations where training of operators can be managed and enforced. This is different from automotive practice, where the operators are members of the general public. Owner's manuals are common, but owner training is not enforced as a common automotive practice. Maintenance training is enforced in VM-managed facilities. In IEC 61508-2, the measures recommended for avoiding systematic errors in specification are almost completely consistent with common automotive practices. The use of formal methods is not common. There is also consistency between the recommendations for controlling systematic faults in design and development, as well as system integration and validation. The recommended techniques used in automotive may also include design, development, and integration of the entire automotive system, including non-safety-related components. The description of practices for high effectiveness is achieved by automotive practices. The broad exception is remote safety systems; co-located safety systems are common.

The performance-testing guidance in IEC 61508-2 is largely applicable to some automotive systems, but not enough. Automotive practice is not only to stress-test systems, but also to test system performance for SOTIF when applicable. This includes discovering situations that are not well-enough defined to predetermine the test conditions to determine performance. This is especially true of region 3 of ISO PAS 21448. SOTIF performance testing is not envisioned by IEC 61508-2. IEC 61508-2 considers functional safety of safety systems – but automotive practice considers more.

IEC 61508-2 provides guidance for performing an SFF. Unlike automotive practices, the SFF considers electromechanical failures and mechanical failure according to IEC 61508.

Automotive may not consider such failures when considering metrics for hardware failures. These failures are considered by automotive practices using other analyses, and measures are taken to ensure safety. IEC 61508-2 specifies other exclusions from the SFF, whereas automotive practice is to include only the components involved in achieving the safety goals. Both include essentially the same electrical parts and preclude gaming the metric.

Specific, detailed guidance for a safety manual is provided by IEC 61508. The purpose is stated to be to allow an item compliant to IEC 61508 to be integrated safely into another system. This is also consistent with automotive practices. Automotive suppliers may also provide safety manuals to customers that detail the functions provided, failure modes and rates sufficiently for the customers to perform needed safety analyses, as well as constraints for safe integration, such as external safety measures assumed. Further, in both IEC 61508-2 and automotive practices, the capability, including systematic capability, is provided with reference to the supporting analyses and confirmation measures. Automotive safety manual practices support similar purposes. IEC 61508 guidance supports these purposes.

IEC 61508-2 provides detailed analysis for implementing on-chip redundancy. There is guidance requiring DFA for such an implementation, and this is common automotive practice as well. Guidance for effective measures such as separate blocks with independent power, non-coupling connections, internal versus external watchdog circuitry, and other methods are considered by both IEC 61508 and automotive practice. However, IEC 61508-2 puts emphasis on determining a β-factor for evaluating independence, and this is not consistent with common automotive practice. ISO 26262 contains a note not recommending the use of a β-factor for quantifying coupling. ISO 26262 provides extensive DFA guidance and examples, and DFA evaluations are used in automotive practice; this is a significant difference.

IEC 61508-3 – Software Requirements

The requirements of IEC 61508-1 and -2 apply to software, as well. Further clarifications, requirements, and considerations for software are provided by IEC 61508-3. Planning must include procurement, development, verification, integration, and modification of software, according to IEC 61508-3. These are major areas of planning in automotive practice. IEC 61508-3 requires configuration management throughout all phases, verification and change control throughout all phases, and controls to ensure that all necessary steps have been completed. There are specific additional requirements for software configuration management. All these requirements are consistent with automotive practices. Engineering quality control is common for managing software development and implementation in automotive; such procedures are differentiators in automotive business awards. There is a strong software relationship between automotive suppliers and VMs. Each may develop software for the other – the software process is key.

The discussions of the phases of a typical software lifecycle provided by IEC 61508-3 are applicable in total for automotive practice. IEC 61508-3 contains extensive normative considerations for requirements. External system interface requirements, hardware safety requirements on software, and internal software monitoring requirements are all considered in the software design requirements. Automotive practice also addresses these sources of requirements. IEC 61508-3 also addresses software validation, which must include validation in the system. Automotive practice also requires in-vehicle validation. There are many significant detailed requirements in IEC 61508-3 for software validation, and they are generally consistent with automotive practices. SOTIF software validation may vary.

IEC 61508 sets objectives for software design and development that are consistent with automotive practices. The context of the design and development for IEC 61508-2 may appear to be industrial rather than automotive, but the concept of a supplier, the VM, or both providing software is applicable to automotive practices. The requirements for software design and development in IEC 61508-3 are consistent with automotive practices, including using parallel structures to improve SIL capability. In automotive, such decomposition is performed on requirements; the parallel structure supports it. IEC 61508-3 requires a software safety manual. Automotive practice may require a safety manual if the software is developed out of the system context; when developed internally, the requirements are linked. This differs from IEC 61508-3.

IEC 61508-3 has requirements for configurable software. While the requirements are applicable to automotive practices, automotive practices consider more detailed requirements for configurable software, as found in ISO 26262. Likewise, IEC 61508-3 has software architecture requirements, which are consistent with automotive practices. Still, automotive has prescriptive consideration in ISO 26262 for software architecture. IEC 61508-3 has requirements for software tools, and these are different from automotive practice. The principles are slightly different, and the definitions of categories are defined differently, based on whether the tool affects the code, fails to detect an error, or could alter the code. In automotive practice, the categories have to do with causing a dangerous failure and vary with ASIL. If the development process can catch the error, tool qualification is not required; but IEC 61508-3 requires tool qualification. Still, the effect of tools is considered by both. IEC 61508-3 has requirements regarding the programming language and the use of guidelines, and so does automotive practice.

IEC 61508-3 has requirements for software design and development. These are consistent with automotive requirements, such as a development interface agreement for joint development. However, automotive practice has further requirements. IEC 61508-3 has requirements for code implementation that specify a level of recommendation for various techniques based on the level of integrity required. Code review of each module is required by IEC 61508-3; this is similar to automotive practices, except that techniques that are more appropriate for industrial applications are not as highly recommended in automotive practice, such as one computer monitoring another in an industrial environment. Automotive may instead use a monitoring processor or another architecture to ensure integrity. IEC 61508-3 requires module testing to a specification. All requirements are to be tested, but equivalence classes, boundary-value testing, and formal methods can be used to reduce the scope. Documentation of the results is required. This is very similar to automotive practice; often, automotive EQA audits this. IEC 61508-3 requires software integration testing to a specification covering all requirements; equivalence classes, boundary value testing, and formal methods can be used to reduce the scope. Documentation is required of the results. Any changes made during integration testing require an impact analysis and change control.

This is similar to automotive practice, and the impact analysis is included in the change-control process. Tools are often used in automotive; the flow is then controlled.

IEC 61508-3 has requirements for software and hardware integration to ensure compatibility and to ensure that the requirements of the SIL are met when integrated. An integration test specification is required that splits the system into integration levels and includes all the information to conduct the tests, including tools, environment, and test cases. Any needed use of the customer site is specified. Failures are dealt with, and design changes require an impact analysis. This differs from automotive practices with respect to specifying the use of the customer site for hardware and software integration, as is the case in an industrial application. Otherwise, it is similar to automotive practice. IEC 61508-3 requires modification procedures after the software is deployed. Automotive is similar but has different procedures since the software is in use by the general public: modification is on the next release, over the air, or by recall. IEC 61508 does not discuss this. IEC 61508-3 discusses the software aspects of validation, and documentation of the results is required. Validation requirements completed during verification are taken into account, and the version of software validated is specified. Failures may require repetition of some development phases. Automotive practice is to complete system validation on the target vehicle with a specified version of software; for SOTIF, all the test cases are not completely specified but are sought systematically in area 3. This is not anticipated exactly in IEC 61508.

IEC 61508 has requirements for changes to software after validation to maintain its safety. A procedure must be available prior to the change. A change is warranted by a safety issue, by a change to the intended function, or because it is discovered that the software is not meeting its requirements. This is similar to automotive practice for changes after launch. IEC 61508-3 requires an impact analysis of the effect of the proposed change on the functional safety of the system, considering known potential hazards and the need to determine whether any new potential hazards have been introduced. This is similar to automotive practices, except that SOTIF may also need to be considered. IEC 61508-3 requires that changes be planned, competent staff be allocated to make the change, and appropriate reverification and revalidation be performed, passed, and documented. The same is required by automotive practices and may include SOTIF.

IEC 61508-3 requires verification of the software lifecycle phase consistent with its inputs and outputs. This is similar to ISO 26262 requirements for normative inputs and work products for software development. In addition, IEC 61508-3 requires verification that the architecture is consistent with the design, the test results are consistent with requirements, and other prescriptive requirements such as timing. These are attributes consistent with the auditing and assessment of work products by system safety personnel and engineering quality personnel. Some of this verification is performed by the design team or contracted for externally, as software development is typically the most labor-intensive part of automotive product development. Customers often engage third-party audits; such process verification is expected. IEC 61508-3 also requires a safety assessment of the software aspects of the safety system, including closure of issues. The same is true for automotive safety practices.

IEC 61508-3 recommends measures and techniques for requirements traceability that agree with automotive practices. IEC 61508-3 recommends measures and techniques for software architecture design that include the high recommendation of a physically separate computer monitoring a controlled computer for SIL 4. While automotive practice often includes a separate processor and memory, with no uncontrolled dependent failures, a physically separate standalone computer for monitoring is not common practice. The other techniques and measures for software architecture are consistent with automotive practices, as are IEC 61508 recommended techniques and requirements for tools. Automotive practices may sometimes require qualified tools; this can be different from certified tools.

IEC 61508-3 recommends measures and techniques for detailed design, module testing, software-to-software integration, and software-to-hardware integration. While these measures and methods are consistent with automotive practices, notably missing from IEC 61508-3 are recommended methods and measures for software unit testing, a common automotive practice in software design. Unit testing may use some similar methods, but others are unique to unit testing. IEC 61508-3 also recommends methods and techniques for software validation, modification, and verification. These are consistent with automotive practices, but none specifically address SOTIF.

IEC 61508-3 recommends some measures and techniques to perform a functional safety assessment of software aspects; these are consistent with automotive practices. Other, more detailed analysis is used in automotive functional safety analyses of software depending upon the available resources of the platform used. These may include fulfilling assumptions of the platform in the requirements and implementation of the software as well as other software safety analyses. These are often checked in automotive practice, but IEC 61508-3 does not mention them.

IEC 61508-3 provides additional detailed tables for software, including tables of methods and measures for detailed code design and coding standards. These are like the coding guidelines used in automotive practice. The dynamic analysis and testing detailed recommendations in IEC 61508-3 are consistent with automotive practice, as are the functional and black box testing recommendations. The software failure analysis recommendations from IEC 61508 are also similar to recommendations considered in automotive practices, as are the modeling recommendations. Many of the IEC 61508-3 recommendations for performance testing, semi-formal methods, and static analysis are considered in automotive practice. Automotive practice is also consistent with IEC 61508-3 recommendations for a modular approach. Sometimes automotive practice uses third-party reviews, which may consider similar recommendations.

IEC 61508-3 provides recommendations and guidance concerning ranking the capabilities of the software system's various techniques and recommendations based on the rigor to be applied to achieve the required systematic capability based on SIL. As in automotive practice, the techniques are not prescriptive, and other techniques can be applied as long as the goals of the phase involved are met. Application of such explicit recommendations may not always be common automotive practice in all areas. Automotive practice is to follow guidance to achieve software systematic capabilities; such guidance is widely measured, and EQA audits this regularly.

Techniques and measures to achieve systematic capabilities for a software safety requirements specification are ranked by IEC 61508-3 based on completeness, correctness, freedom from specification faults, understandability, freedom from interference, and providing a basis for verification and validation. These properties may not specifically be evaluated in automotive when considering how to generate a software specification. Nevertheless, the techniques listed are considered for appropriateness based on the application. The rankings are consistent with automotive considerations, and an appropriate combination can be chosen.

IEC 61508-3 ranks techniques and measures to achieve systematic capabilities for a software architecture design based on completeness, correctness, freedom from design faults, understandability, predictability, verifiability, fault tolerance, and robustness against dependent failures. These properties are evaluated in automotive when considering how good an architecture is for an application. The rankings are consistent with automotive considerations, but stateless design, artificial intelligence fault correction, and dynamic reconfiguration are not considered in automotive practices. The other techniques are common, and an appropriate combination is chosen.

Techniques and measures to achieve systematic capabilities for software support tools and programming language are ranked by IEC 61508-3 based on production support, clarity of operation, and correctness. Automotive practice is to use a more systematic process to classify software tools. Qualification is necessary for some tools, and other process considerations are taken into account.

IEC 61508-3 ranks techniques and measures to achieve systematic capabilities for detailed software design based on completeness, correctness, freedom from design faults, understandability, predictability, verifiability, fault tolerance, and robustness against dependent failures. These properties are evaluated in automotive when considering how good a detailed design is for an application. The rankings are consistent with automotive considerations, but many other considerations are also common in automotive practices. An appropriate combination is chosen, and third parties may evaluate practices.

Techniques and measures to achieve systematic capabilities for software module testing and integration as well as performance testing and validation are ranked by IEC 61508-3 based on completeness, correctness, repeatability, and precision in defining test configurations. These properties are evaluated in automotive when considering how to verify and test software modules and how to execute performance testing. However, with respect to validation, process simulation is not considered, as it applies to validating a safety system for use in a facility. The other rankings are consistent with automotive considerations; an appropriate combination is chosen. Automotive practice may also consider SOTIF validation, whereas IEC 61508-3 does not address SOTIF.

IEC 61508-3 ranks techniques and measures to achieve systematic capability for software modification based on completeness, correctness, freedom from design faults, avoiding unwanted behavior, verifiability, and regression testing. These properties are evaluated in automotive when considering implementing a change in an application. The rankings are consistent with automotive considerations; an appropriate combination is chosen, and EQA may audit these choices when released.

Techniques and measures to achieve systematic capabilities for software module verification are ranked by IEC 61508-3 based on completeness, correctness, repeatability, and precision in defining test configurations. These properties are evaluated in automotive when considering how to verify and test software. The rankings are consistent with automotive considerations, and an appropriate combination is chosen. Automotive practice may also consider SOTIF verification, whereas IEC 61508-3 does not address SOTIF.

IEC 61508-3 ranks techniques and measures to achieve systematic capabilities for software functional safety assessment based on completeness, correctness, issue closure, modifiability, repeatability, timeliness, and precision of configuration identification. These properties are similar to those evaluated in automotive when considering developing an overall functional or system safety assessment process, though perhaps not systematically. These rankings are consistent with automotive considerations; an appropriate combination is chosen, and SOTIF considerations can be added.

Detailed tables are also provided in IEC 61508-3 that rank techniques and measures based on design and coding standards, dynamic analysis and testing, functional and black-box testing, failure analysis, modeling, performance testing, semi-formal methods, static analysis, and modular approach. The rankings are for the techniques' effectiveness in appropriate areas. All the techniques are appropriate for automotive, though automotive may not select them explicitly with respect to these rankings. Nevertheless, such selection is implicit even if less systematic. The rankings are valid for automotive practices, and the areas ranked are appropriate.

IEC 61508-3 provides additional requirements for the safety manual for software elements. These are in addition to the requirements provided for safety manuals in general: in particular, the runtime environment, recommended configuration of the software element, hardware restrictions, competence of the integrator (such as knowledge of specific software tools), and other assumptions that are required to be fulfilled for use. This is consistent with automotive practices. There are many other IEC 61508 requirements for the safety manual for a compliant software element. Not every automotive safety manual may need to meet all of them, though most are applicable. They are consistent with automotive practices and may be considerations.

IEC 61508-3 provides guidance for which clauses of IEC 61508-2 are applicable to software. The considerations that are common, such as integration, verification, and validation, may also be considerations in automotive practice. However, standards used in automotive in general may not require software domain specialists to mine applicable requirements from the hardware lifecycle. Considerations common to all lifecycle phases can be combined, such as documentation and normative considerations for requirements. This guidance applies to hardware and software; combining the guidance reduces repetition.

Guidance concerning techniques for achieving non-interference between software elements on a single computer is provided by IEC 61508-3. This guidance discusses spatial and temporal domains of interference as data and timing, respectively. These domains are applicable to automotive practice. Encapsulation and loose coupling are techniques that are encouraged, while global coupling and content coupling are discouraged. The terms and advice are directly applicable to automotive practice. Automotive applications typically use one computer, and further considerations may also apply.

IEC 61508-3 provides guidance for tailoring lifecycles associated with data-driven systems. The application and data-driven parts of the software are defined and guidance is given to determine how much of the safety lifecycle is relevant for different configurability examples. This type of determination is appropriate for software in a manufacturing scenario but is not common automotive practice. There is guidance for calibration in automotive systems that utilize equivalence classes; otherwise, an impact analysis is commonly used to determine applicable lifecycle phases.

IEC 61508-4 – Definitions and Abbreviations

IEC 61508-4 provides terms and definitions that were often referenced in automotive practice prior to the release of ISO 26262. In developing ISO 26262, clarifications were made in the definitions to support the automotive context. Some definitions and abbreviations were added, some were deleted, and some definitions were modified in support of automotive needs. As many automotive safety practitioners also have a background in safety that used the definitions and abbreviations of IEC 61508-4, understanding the differences is helpful. The definitions of ISO 26262 are referenced in ISO PAS 21448, and the differences may be significant.

Some abbreviations in IEC 61508-4 are not normally referenced in automotive practice. These include *as low as reasonably practicable* (ALARP), the practice of further reducing a risk beyond what is required for compliance where practical, though this practice may still be used in some automotive applications. Other abbreviations that are not commonly used are complex programmable logic device (CPLD), electrical/electronic/programmable electronic system (E/E/PE) equipment under control (EUC), generic array logic(GAL), hardware fault tolerance (HFT), M out of N channel architecture (MooN), mean time to repair (MTTR), and mean repair time (MRT) (some reliability experts who work in automotive may use these, but they are not as common among automotive safety experts), as well as programmable logic controller (PLC), programmable logic sequencer (PLS), programmable macro logic (PML), safe failure fraction (SFF), safety integrity level (SIL), and very-high-speed integrated circuit hardware description language (VHDL). The reasons that these IEC 61508-4 abbreviations are not common in automotive are twofold. First, many of the abbreviation in IEC 61508-4 target industrial applications where these abbreviations are common, such as a PLC. Second, some of the definitions in IEC 61508-4 are included in support of the IEC 61508 standard itself. Some of the abbreviations listed in IEC 61508-4 that are commonly used in automotive practice are not listed in ISO 26262. So, for someone coming into automotive safety from another industry, or someone who used IEC 61508 for automotive safety work before other standards became available (like this author), this list is useful for ease of communication. Automotive practice is to use many abbreviations, so the list can help avoid confusion.

There are significant differences in the safety terms in IEC 61508-4 compared to automotive practice. IEC 61508-4 includes damage to property in the definition of *harm*; automotive practice only includes injury or damage to the health of persons. IEC 61508-4 defines a *hazardous event* as an event that may result in harm; automotive practice is to only include vehicle scenarios. IEC 61508-4 defines *tolerable risk* in a way that is the opposite of *unreasonable risk* in automotive practice.

Safety is defined as the absence of unacceptable risk in IEC 61508-4; ISO 26262 uses the absence of unreasonable risk. In the committee discussion, this was done in deference to German legal practices, where it was said that the judge decides what is acceptable.

The *safe state* in IEC 61508-4 is defined as the state of the equipment under control achieving safety. In automotive practice, a *safe state* is the operating mode achieved after a functional failure without unreasonable risk. The major difference is that automotive practice is to only discuss a safe state in the context of recovering from a failure. In SOTIF this is discussed as a minimum risk condition.

IEC 61508-4 defines a *functional unit* so that it can be hardware or software. In automotive practice, it is usually distinguished either as a *hardware part* or as a *software unit*. *Architecture* is defined in IEC 61508-4 as a specific configuration of hardware or software, where automotive practice also includes in the definition the requirements assigned to each element of the configuration.

IEC 61508-4 defines an *element* as part of a subsystem that performs a safety function. Automotive practice is to consider an *element* in the context of an item, where an item is a system or system of systems at the vehicle level. So, when an item is a system of systems, an element can be a system. Likewise, when an item is not developed in the context of a vehicle but is intended to be integrated into a vehicle later, the item is referred to as a *safety element out of context* (SEooC). Thus, *element* has a significantly different meaning in automotive safety.

Fault has essentially the same meaning in IEC 61508-4 as *fault* does in automotive practice. IEC 61508 defines *fault* in the context of a functional unit, whereas automotive uses the term *fault* in the context of an item or element. The same is true of *failure*.

IEC 61508-4 defines *process safety time* as the time available to prevent a failure from leading to a hazardous event. This is similar to the term *fault tolerant time interval (FTTI)*, commonly used in automotive practice to define the same interval. FTTI includes the *fault detection time interval* and the *fault reaction time interval* (taken together, these become the *fault handling time interval*) as well as any *emergency operation time interval* that is needed before reaching a permanent safe state. The FTTI is much more prescriptive and is defined at the item level because hazards are only at the vehicle level. There was considerable discussion in the ISO 26262 committee about this, as sometimes FTTI was used at the component level in automotive practice, despite the definition and supporting logic concerning hazards. The definition and supporting figure were finally agreed on, and the definition references the figure.

IEC 61508-4 defines *mean time to restoration* (MTTR), which is not a term widely used in automotive practice. The definitions of *safety lifecycle* and *software lifecycle* in IEC 61508-4 agree with common usage in automotive practice. IEC 61508-4 notes that *validation* is performed in three phases: system, electronics, and software. Automotive practice is to consider validation at the vehicle level for the item. Otherwise, the terms defined in IEC 61508-4 are similar to the understanding in automotive practice. Some of the terms are defined for use in the IEC 61508 and are not used in automotive practice; these are not discussed here.

IEC 61508-5 – Examples of Methods for the Determination of SILs

IEC 61508-5 provides generalized examples of ways to determine SILs. The relationship of risk to SIL is explained so that the necessary requirements of IEC 61508 parts 1, 2, and 3 can be determined. Though the specific methods and calibrations are different, the relationship of risk to ASIL in ISO 26262 is somewhat similar. Automotive practice generally does not compute a numeric value of risk reduction when determining an ASIL. Whereas IEC 61508-5 discusses a necessary risk reduction, the ISO committee rejected this concept early in the development of ISO 26262. The target risk for employees versus the general public discussed in IEC 51508-5 is not normally considered applicable in automotive practice, where the risk to road users is considered. While agencies of the government may consider societal risk in policy decisions, as discussed in IEC 61508-5, this is not common practice in evaluating the risks of hazard for an individual automotive application. IEC 61509-5 is considering societal risk in an industrial or utility hazard context, and the role of the safety system in reducing the risk of the equipment under control. Automotive practice is to consider the development of the function and the safety mechanisms together.

Consideration of safety integrity is very similar between IEC 61508-5 and automotive practice. Hardware safety integrity and systematic safety integrity are covered by both IEC 61508-5 and automotive practices for functional safety. IEC 61508-5 considers low-demand mode, high-demand mode, and continuous mode, whereas automotive practice may reflect simply on-demand mode or continuous mode. The models for risk reduction for each of these modes are not commonly used in automotive practice, though the concept of achieving an acceptable level of risk is considered, particularly in SOTIF. Common cause is discussed in IEC 61508-5, and the discussion is also relevant for automotive practice, especially when decomposition of requirements with respect to ASIL is employed, or in SOTIF applications where redundant sensors may have common limitations. The same is true in the case of multiple layers, as discussed in IEC 61508-3. The discussion of the difference between risk and SIL is also consistent with automotive considerations, such as ASIL. The discussions in IEC 61508-5 of SILs and systematic capabilities of software are consistent with automotive understanding. The discussion in IEC 61508-5 of the allocation of safety requirements to other technologies is somewhat different from automotive practice because of the inclusion of a numerically quantified risk-reduction allocation, but otherwise it is consistent. IEC 61508-5 also discusses mitigation systems that are not directly discussed in automotive practice. However, SAE J2980 does discuss the use of traffic statistics in the determination of ASILs; these statistics include the effects of automotive mitigation systems, such as airbags and seatbelts. The severity or exposure is reduced in these statistics due to the effects of these mitigation systems, but this reduction is not quantified specifically.

IEC 61508-5 discusses multiple ways to determine the SIL of the requirement to avoid a hazard. The guidance suggests multiple methods to be used first for screening conservatively and then another method for more precise assessment if the SIL determined in the initial screening is too high. Automotive practice does not use multiple methods to determine an ASIL but uses various approaches to determine the parameters used to determine the ASIL. These range from qualitative consensus methods, applying guidance from ISO 26262, to more precise methods that use traffic statistics. Still, the ASIL is not determined quantitatively; the parameters have quantitative definitions.

The theory of ALARP and tolerable risk concepts are explained for application in SIL determination. IEC 61508-5 discusses a region of intolerable risk, a region of generally tolerable risk, and a region in between that may be tolerable based on reducing the risk ALARP. In this region, a risk is accepted if the cost of reducing it further is considered impractical. This may result in a lower SIL for risk reduction. Automotive practice does not use such an evaluation for determining ASIL or in SOTIF applications. ALARP is referenced if an ASIL is determined and additional measures are practical. Then the ALARP principle may recommend such measures, and the risk is further reduced.

Prior to ISO 26262 being widely used in automotive practice, the determination of the required risk reduction for SIL classification was used. For example, consider determining the SIL for avoiding autosteer in an electric steering system that is to be deployed in a hypothetical country with good infrastructure and unlimited speed on the autobahn. The death rate given an accident is no more than 1 death in 200 accidents. The acceptable user risk is a 10E-9 probability of death. Then the required risk reduction could be 2E-7 to obtain this acceptable user risk, which corresponds to SIL 2 (10E-6 to 10E-7). Now, similar traffic statistics are used to determine severity or exposure, as discussed in SAE J2980. It is no longer used for ASIL – it is used for SOTIF. ISO PAS 21448 discusses this.

The risk graph approach explained in IEC 61508-5 is similar to the method used in automotive practice. Severity in IEC 61508-5 is based on anything from a minor injury to many people killed, as in a potential explosion in a nuclear plant. Automotive practice considers the severity of the injury, up to probable death, of only one person. IEC 61508-5 discusses frequency of exposure or exposure time as either high or low, while automotive has four categories each of frequency and time and an additional "incredible" category (E0) for both. IEC 61508-5 has two categories for controllability or ability to avoid harm, while ISO 26262 has three plus "controllable in general." IEC 61508-5 has a probability of the unwanted occurrence parameter that is a qualitative indication of the likelihood of the harm, given the hazardous event. In automotive practice, this is like the probability of death given an accident. For systems like steering and braking, this takes into account passive safety systems, for example. Early in the discussions of the ISO 26262 committee, but after the bilateral discussions between France and Germany, this parameter, *W*, was discussed because it was not included in the first proposed draft. A replacement parameter, *likelihood*, was proposed by the chairman of the US delegation. It was agreed in principle to be potentially applicable but was rejected because it was thought that it might reduce the ASIL below D of potential hazards that can lead to accidents where death is possible, though not likely. This could therefore be potentially interpreted as not requiring state-of-the-art methods and measures. Later, it was noted that such statistics are taken into account in the distribution of injuries, potentially affecting severity based on the likelihood of serious injuries or death (or alternatively exposure, though not explicitly stated in ISO 26262). Nevertheless, the ASIL determination never added *likelihood*; it is considered indirectly.

IEC 61508-5 discusses a semi-quantitative method using layer of protection analysis (LOPA). An example is given using an industrial scenario. This has not generally been applied to automotive product implementations; the quantification of the intermediate steps does not have a direct analogy, and the resulting quantification is not applicable to ASIL determination. Even in SOTIF applications, the use of LOPA is not common practice. IEC 61508-5 also discusses determination of SILs using a qualitative method with a hazardous-event severity matrix. It is for use when the quantitative data is not available and reduces the likelihood of the hazard by an order of magnitude for each independent risk-reduction method used. Automotive practice similarly reduces the ASIL if another system prevents a hazard, such as an engine management system limiting cruise control authority. Nevertheless, the practice described in IEC 61508-5 for such a qualitative method is not common automotive practice. ASIL determination is not based on likelihood reduction; ISO 26262 practices are followed.

IEC 61508-6 – Guidelines on the Application of IEC 61508-2 and IEC 61508-3

IEC 61508-6 provides guidance in the form of a flow of activities that implement the guidance for hardware and software given in IEC 61508-2 and IEC 61508-3, respectively. A waterfall model is assumed, but it is noted that equivalent models may be used. The flow corresponds to implementation in an industrial application but can be applied to an automotive application. The steps are followed in automotive practice, taking into account the different application; the flow of requirements and review steps are similar.

The first step in the guidance for applying IEC 61508-1 discussed in IEC 61508-6 is determining the allocation of safety requirements to the safety-related system and splitting them between hardware and software. This is also common automotive practice. Then IEC 61508-6 discusses starting the validation planning that determines the hardware architecture and modeling it to see whether it fulfills the failure rate requirements. This does not always happen next in automotive practice, but the requirements linking to verification methods may begin, and the hardware architecture is determined. ISO 26262 does not include a hardware metric until the hardware components are determined in the design. This step was proposed but not accepted. If the hardware is substantially a carryover from a similar product, then evaluating the changes can reduce the risk of waiting until the design is detailed to evaluate the probability of a hardware failure.

Then, IEC 61508-6 suggests that the probability of a hardware failure be estimated; and if this estimate is not achieved then a component should be changed or the architecture fixed. This comes later in automotive practice for modifications but might be attempted for new development. Next, IEC 61508-6 discusses implementing the design, integrating the software with the hardware, validating the hardware and software, integrating them into the overall system, and validating. This is somewhat like automotive practice in that after the system is verified, it is re-verified when production is tooled, a step commonly called *validation*. Safety validation is on the vehicle, which differs from IEC 61508-6.

IEC 61508-6 guidance applying IEC 61508-3 is similar to that just discussed. First, get the requirements and determine the software architecture, and review it with the hardware developers. Then, start planning the verification and validation, and implement the design. All the appropriate measures and techniques contained in IEC 61508-3 are to be applied. Integrate with the hardware developer, and perform verification and validation. This is somewhat like automotive practice, where safety validation is on the vehicle.

IEC 61508-6 provides an example of a technique for evaluating probabilities of hardware failure that is composed of reliability calculation techniques and tables assessing several potential architectures. The reliability computation techniques are directly applicable to automotive practices, including the use of reliability block diagrams and low- and high-demand cases. The cases that take into account repair and downtime are less likely to be applied in automotive systems, as their primary use case is an industrial application. Proof-testing is discussed, using periods that are common for industrial applications, and the tables are based on these. In automotive practice, proof-testing is like startup or shut-down testing of a system for each ignition cycle and is much more frequent than assumed in IEC 61508-6. This makes the tables more conservative than automotive practice. The use of a β-factor for common cause failures between channels might be used in automotive practice but is not required by ISO 26262. Directly addressing dependent failures is common, and the 1oo1, 1oo2, etc. architectures are considered in automotive for some safety-related systems. The calculations are useful. System downtime is not considered; there may be architecture variations.

IEC 61508-6 discusses the principle of PFD calculations in the context of Petri nets, which can be and have been used in automotive practice. The examples used in IEC 61508-6, while useful in the discussion of Petri nets and PFD applications, may not represent a typical use in automotive practice. The discussion of repair times is more applicable to industrial applications. Nevertheless, using the technique to systematically develop or review algorithms and safety mechanisms in an automotive system context is useful. The use of Petri nets is not discussed in ISO 26262 or ISO PAS 21448, though members of the committees mentioned them. Petri nets may help comply with both; IEC 61508-6 provides a reference.

Other approaches to modelling the safety-related system for PFD calculations are discussed in IEC 61508-6. It is discussed that the size of the model with respect to the size of the system is linear with fault tree analysis (FTA) and Petri nets, making them more popular for larger systems. This is applicable to automotive practice where an FTA or equivalent analyses is common practice. Handling uncertainties is also discussed in IEC 61508-6, including the use of Monte Carlo simulation with the incorporation distributions for uncertain parameters. Multiple runs are discussed, to produce a histogram; this technique is employed in automotive practice, particularly for determining ASIL, and in SOTIF. ISO PAS 21448 provides a reference for a presentation of its use for determining the severity of unintended deceleration. Distributions are used for the human reaction trigger variable of growth rate of the angle subtended by the closing lead vehicle, human reaction time, as well as the following distance. Differential speed at collision is calculated in thousands of simulation runs, severity is estimated based on the differential speed if there is a rear-end collision, and controllability is estimated based on the fraction of runs that involve a collision. This is repeated for different deceleration profiles, and ASILs are then determined.

The worked example for diagnostic coverage and SFF in IEC 61508-6 is approximately the same as determining the single point fault metric (SPFM) in ISO 26262. The tables used for determining diagnostic coverage claimed for different failure modes have been updated in ISO 26262, which only allows consideration of safe failures for parts that are involved in achieving the safety goals. This was a compromise agreed to in order to include safe failures when the SPFM was changed from diagnostic coverage only; the change was made to take into account inherently safe designs. This rule does not change the calculation but was considered important by the ISO 26262 committee so that the SPFM could not theoretically be distorted through the inclusion of low-reliability parts with no apparent safety purpose. Automotive cost pressures may discourage this practice, but nevertheless, the rule was added.

IEC 61508-6 discusses a methodology for quantifying the effect of hardware-related dependent failures in electronic systems. This quantification is for estimating the allowable improvement claimed in multiple-channel systems for dependent failures that have not been resolved by dependent failure analyses, using a β-factor that is systematically determined based on engineering judgment. Using such a factor is not common automotive practice since the introduction of ISO 26262, which does not require it and includes a second method for hardware failure rates that does not require computing the overall failure rate. Nevertheless, a β-factor is estimated in some applications by some developers. The procedures in IEC 61508 for systematically making this estimate are not commonly used in most automotive practice, and the alternative shock method is rarer. Instead, automotive practice is to systematically perform a DFA and rely on this analysis. ISO 26262 provides extensive guidance, and there are references to further tutorial material. Automotive practice also considers software-dependent failures, and tutorials are often available.

Two examples of the application of software safety integrity tables from IEC 61508-3 are provided in IEC 61508-6. Both examples are for industrial applications; one has SIL 2 requirements, and the other has SIL 3 requirements. Automotive practice is to use ISO 26262 as a functional safety reference, so the methods and measures in the tables are not completely applicable. Nevertheless, justifications used in application may be similar to the comments used in an automotive application. Justification of every technique that is recommended but not used is commonly provided in automotive practice; this normally is done systematically and then assessed with the independence required for the ASIL. Sometimes the customer or the customer's auditor also reviews such justifications. So, to this extent, the IEC 61508-6 example is applicable, because automotive practice is similar.

IEC 61508-7 – Overview of Techniques and Measures

IEC 61508-7 includes descriptive overviews of techniques and measures for electronic safety-related systems for controlling random hardware failures. These descriptions are very similar to the techniques discussed in ISO 26262, with some exceptions. IEC 61508-7 includes a discussion of detecting relay failures that is relevant to automotive practice but not included in ISO 26262. ISO 26262 provides a more general description of using a comparator that could include comparing two processing units. The idle current principle (de-energized to trip) and analogue signal-monitoring techniques discussed in IEC 61508-7 are directed to industrial applications and normally not applicable in automotive practice. The IEC 61508-7 techniques of standard test port and boundary scan architecture, monitored redundancy, de-rating, self-test by software (limited number of patterns and walking bit), reciprocal comparison by software, word-saving multi-bit redundancy (for example, ROM monitoring with a modified Hamming code), modified checksum, and electrical/electronic components with automatic check are applicable to automotive practice, although they are named and discussed somewhat differently in ISO 26262. Coded processing of one channel is not common automotive practice. Single-word signature, double-word signature, and block replication are memory-protection techniques in IEC 61508-7 that may also be used in automotive practice. Likewise, the advice in IEC 61508-7 to protect volatile memory from soft errors is applicable to automotive practice. References are provided for the techniques discussed in IEC 61508-7. The techniques suggested in IEC 61508-7 for RAM protection can be applicable for automotive practice, though the use of an error-correcting code (ECC) to protect RAM is common automotive practice.

All the techniques suggested in IEC 61508-7 for data-path protection are relevant in automotive practice. Some are implemented in automotive practice using a cyclic redundancy data extension and a rotating code that are checked by the receiver, as well as a redundant transmission or redundant sources and comparisons. These are common implementations in automotive practice. The IEC 61508-7 discussions of power-supply protection are also applicable in automotive practice. IEC 61508-7 discusses several techniques for logical and temporal program sequencing that are applicable in automotive applications; often, resources are provided by the processor used to support logical and temporal program monitoring.

The monitoring techniques in IEC 61508-7 for ventilation and heating are applicable for automotive practice, though forced air is not common in automotive systems. Thermal sensors are common in automotive practice. The IEC 61508-7 techniques for communication and mass storage target industrial applications. Spatial separation of power and signals is less common in automotive practice when only one system connector is available. The sensor protection discussed in IEC 61508-7 also targets industrial applications; ISO 26262 includes a significantly expanded description of sensor-protection techniques. The monitoring techniques for actuators in IEC 61508-7 are applicable to automotive practice, and cross-monitoring is possible in rare circumstances. IEC 61508-7 also suggests putting electronics in an enclosure designed to protect against the environment; this is common automotive practice.

IEC 61508-7 provides an overview of techniques and measures for safety-related systems for the avoidance of systematic failures. The general measures and techniques for project management and documentation are applicable to automotive practice. The separation of system safety functions from non-safety-related functions discussed in IEC 61508-7 targets industrial applications; there is an analogue in automotive practice for decomposition of requirements with respect to ASIL to achieve a similar purpose. Diverse hardware aims to detect systematic failures during operation of the equipment under control. Here, there is an analogue to some common architectures in automotive where there is a second processor or another IC is used to detect such failures. Sometimes a challenge and response are used. End-to-end protection is also used.

Guidance regarding techniques to provide a specification free from mistakes is given in IEC 61508-7, including methods that are also applicable to automotive practice, such as a structured specification to reduce complexity; and formal methods, though those are not extensively used in automotive. Many examples of semiformal methods are described in IEC 61508-7, most of which are used in automotive, some to a larger degree than others. IEC 61508-7 provides guidance on methods to produce a stable design in compliance with specifications, and all the techniques described are consistent with automotive practice.

Operational and maintenance guidance is provided in IEC 61508-7, including procedures, user friendliness, maintenance friendliness, and limiting operation possibilities by detecting user errors. While this guidance targets an industrial application, there are analogues in automotive practice to avoid misuse and safety-related maintenance mistakes. The discussion in IEC 61508-7 of limiting operation to skilled operators is not assumed by automotive practice except perhaps in commercial applications. The protection against operator mistakes discussion in IEC 61508-7 is applicable in automotive practice, as is modification protection when appropriate. Input acknowledgment is also suggested in IEC 61508-7 to allow the operator to detect errors before operating the equipment. Automotive practice does not have a direct analogue to this, though some protection is gained by arming a system before accepting input settings, as well as messaging.

IEC 61508-7 provides guidance on methods to be used in system integration. The stated objective is to reveal any failures during the integration phase and previous phases and to avoid failures during integration. Functional testing is proposed to avoid failures during integration and implementation of hardware and software, as well as to discover failures in design and specification. The description of the technique includes guidance applicable to automotive practice. Black-box testing is described by IEC 61508-7 to check dynamic behavior. Data that exercises equivalence classes to their limits is discussed, and this technique is consistent with automotive practices. Statistical testing is also described in IEC 61508-7 and is generally applicable to automotive practice. The field experience description in IEC 61508-7 dismisses reuse of safety mechanisms that have not actually detected problems in the field; automotive practice is to reuse such safety mechanisms that have not caused problems and have not missed detecting a failure (such failures are rare).

System safety validation to prove that the safety-related system meets its requirements, including environmental and reliability requirements, is described in IEC 61508-7. The techniques include functional testing under environmental conditions, surge and electromagnetic interference, static analysis, and dynamic analysis and testing. These validation techniques are consistent with automotive practices. Failure modes and effects analysis (FMEA), cause-consequence diagrams, event tree analysis, failure criticality analysis, fault tree models, Markov models, Monte Carlo simulation, generalized stochastic Petri net (GSPN) models, worst-case analysis and worst-case testing, expanded functional testing, and fault-insertion testing are also described by IEC 61508-7 as validation techniques and are consistent with automotive practices but are normally performed before automotive validation. Markov models and GSPNs are not as commonly used in automotive as the others. An FMEA is rarely omitted for a safety-related automotive system; it is performed early in the design and then updated systematically.

IEC 61508-7 provides an overview of techniques and measures for achieving software safety integrity, including structured diagrammatic methods. The description is consistent with automotive software design practices. IEC 61508-7 describes three other design techniques related to requirements elicitation and capture: controlled requirements expression (CORE), Jackson system development (JSD), and real-time Yourdon. Automotive practice typically employs SPICE [16], and automotive processes are developed based on this reference model. Other schemes are referenced or incorporated, such as real-time Yourdon; however, SPICE audits are common. Data-flow diagrams are described in IEC 61508-7 and are commonly used in automotive practice. Structure diagrams, formal methods, calculus of communicating systems (CCS), communicating sequential processes (CSP), higher-order logic (HOL), language for temporal ordering specification (LOTOS), OBJ (an algebraic specification language), temporal logic, the Vienna development method, and Z are described in IEC 61508-7. These techniques are employed by some automotive software specialists for an application but have not been adopted across automotive software processes. Defensive programming, described in ISO 26262-7, is a broadly adopted automotive practice.

Design and coding standards are discussed in IEC 61508-7. Coding standard suggestions are provided to support the requirements and recommendations in IEC 61508-3, and these recommendations are consistent with automotive practice. IEC 61508-7 describes the rule of no dynamic variables or dynamic objects in order to be able to check memory at startup.

Otherwise, it is recommended to perform online checking of dynamic variables. Limited use of interrupts is described to support testability; these descriptions are consistent with automotive practices, as is limited use of pointers. IEC 61508-7 recommends limited use of recursion, while ISO 26262 recommends no recursion. IEC 61508-7 describes structured programming to reduce complexity, along with use of a modular approach, encapsulation, and trusted software elements, consistent with automotive practice. PIU is described in IEC 61508-7 for software elements; however, the standard for PIU is 10 times higher in automotive practice. IEC 61508-7 requires evidence such as a safety manual to determine whether pre-existing software is compliant; this is consistent with automotive practice. Traceability between requirements, design, and verification is described in IEC 61508-7 and is consistent with automotive practice. Stateless design, as described in IEC 61508-7, is not common practice in automotive because of the complexity of safe behavior for the automobile: for example, taking into consideration inertial sensor inputs for a stability control system. In IEC 61508-7, the description of offline numerical analysis is not a practical consideration in automotive practice. Message sequence charts, described in IEC 61508-7, are employed by some automotive software specialists for an application but have not been adopted across automotive software processes.

IEC 61508-7 describes techniques that are used in software architecture design, including fault detection and diagnosis, error detecting and correcting codes, failure assertion programming, use of diverse monitoring, software diversity, backward recovery, re-try fault mechanisms, and graceful degradation; all of these are consistent with automotive practice, though backward recovery is rare. The IEC 61508-7 descriptions of artificial intelligence fault correction and dynamic reconfiguration are not consistent with automotive practice. The use of time-triggered architectures, UML, class diagrams, use cases, and activity diagrams are consistent with automotive practice.

IEC 61508-7 describes development tools and programming languages, including strongly typed programming languages and language subsets, which are consistent with automotive practice. It describes the use of certified tools and certified translators, and the use of tools and translators that gain increased confidence from use; this is consistent with automotive practice, though the use of the process to detect errors may reduce the need for certification. Comparison of the source program to the executable code is described in IEC 61508-7 to mitigate the need for qualified compilers. While this is not common automotive practice, a similar advantage is possible in model-based development with a qualified modeling tool by comparing the results of the model-generated test vectors of the model with the results of the integrated software using the same test vectors. The use of suitable programming languages described in IEC 61508-7, automatic software generation, and the use of test management and automation tools is consistent with automotive practice, though the list of languages recommended for different SILs is not applicable.

IEC 61508-7 describes probabilistic testing to determine software reliability. This is performed in automotive practice for some product introductions; but for practical reasons, it is done rarely and not for every application. In addition, field data becomes available and can be used. Data recording and analysis, interface testing, boundary value analysis, error guessing, error seeding, equivalence classes and input partition testing, structure-based testing, control-flow analysis, and data-flow analysis, all of which are described in IEC 61508-7, are common automotive practice. Symbolic proof, formal proof, and model checking are also described in IEC 61508-7 but are not common automotive practice. IEC 61508-7 describes complexity metrics, formal inspections, walk-throughs, design reviews, prototyping, and animation that are consistent with automotive practice. Process simulation is described in IEC 61508-7 targeting industrial applications and is not consistent with automotive practice.

Performance requirements are described in IEC 61508-7 that are applicable to automotive practice when traceability is considered. The descriptions of performance modelling, stress testing, response timing, and memory constraints in IEC 61508-7 are consistent with automotive practice. IEC 61508-7 describes an impact analysis of software changes with respect to the effects on other modules and criticality. While this is consistent with automotive practices, with respect to functional safety, the safety-related changes are evaluated for potential effects on safety requirements. The descriptions of the resulting verification are consistent with automotive practice, and updating previous phases of the safety lifecycle may also be considered. Software configuration management and regression validation, described in IEC 61508-7, are consistent with automotive practices. Animation of the specification of a design is described in IEC 61508-7 in a manner more applicable to an industrial application than to automotive practice. Model-based testing, as described in IEC 61508-7, also is applicable to an industrial application and may include a model of the equipment under control. There are some analogues to automotive practice: model-based testing may consider software requirements and the architecture as well as test coverage, particularly if an executable model is used. The model enables the use of back-to-back tests, as recommended in ISO 26262.

ISO 26262-7 discusses functional safety assessments for software. Truth tables are described in a way consistent with automotive practice. The HAZOP is described in a way that is appropriate for industrial application or for system hazard analysis performed in automotive practice in the concept phase and not normally repeated during software safety assessment. A HAZOP technique can be applied to the software architecture to elicit the effects of failures; this is different from the IEC 61508-7 description. The description of common-cause failure analysis in IEC 61508-7 contains parts that target industrial applications and parts that are applicable to automotive practice. In ISO 26262, the guidance on common-cause analysis is included in the guidance on dependent failures. Reliability block diagrams are described for software in IEC 61508-7, but this description is not consistent with automotive practice.

IEC 61508-7 provides a description of a probabilistic approach to determining software safety integrity for pre-developed software. Mathematical considerations are explained. While the principles are applicable to automotive practice for PIU considerations, the criteria are not applicable.

An overview of techniques and measures for design of ASICs is provided in IEC 61508-7. The descriptions contain recommendations that are specific to integrated circuits, including use of design descriptions at a high functional level, schematic entry, structured descriptions, proven tools, simulation, module testing, and high-level testing of the ASIC. These considerations are applicable to automotive practice. There are also many general descriptions of specific design rules and coding standards for ASICs, which are also considered in automotive practice. Testing, using a PIU production process, and quality control are described and are considered in automotive practice.

Definitions of the properties of a software lifecycle process are provided by IEC 61508-7, with respect to completeness, correctness, consistency, understandability, freedom from unnecessary requirements, and testability of the requirements. These are also considered in automotive practice. Definitions of software architecture design properties are provided by IEC 61508-7 and include completeness and correctness of addressing the requirements, as well as freedom from design faults, predictability, fault tolerance, and testability. These are also consistent with automotive considerations. Properties of software support tools, programming languages, and detailed design definitions provided by IEC 61508-7 are consistent with automotive considerations. The definitions of properties provided for software module testing and integration, integration of software with programmable hardware, verification and safety validation, software modification, and safety assessment by IEC 61508-7 are consistent with automotive practices. IEC 61508-7 provides guidance for the development of safety-related object-oriented software; such guidance is not provided in ISO 26262 specifically for object-oriented software, but object-oriented software is common in automotive practice. The IEC 61508 guidance is considered.

Appendix B

ISO 26262 – Notes on Automotive Implementation

Introduction

ISO 26262 was originally released November 15, 2011, and has been revised December, 2018. One of the reasons for the revision was to update it based on usage. Because the standard was written so that many experts could agree on the wording, different interpretations and practices are used with the intent of complying with the standard. The statements and judgements made in this appendix represent the opinion of the author based on his experience. They do not represent the results of a survey of the industry and are not intended to be absolute. Nevertheless, for personnel who may have had experience in other industries, this review may be useful. ISO 26262 addresses functional safety only; it does not address safety of the intended function (SOTIF).

Also, the definitions are not grouped as they are in IEC 61508-4 – they are grouped in this discussion in order to preserve some context. In addition, there are pages of acronyms for use in the standard. Many are common terms for electrical engineers working in the automotive domain, but others are not, like *BB* for *body builder* (a T&B term). This can make reading the standard more cryptic. If the reader is purchasing parts of the standard, part 1 is advisable; the glossary contains much informative content and will be useful.

ISO 26262-1 – Glossary

Use of Terms

There are 185 definitions in ISO 26262-1, more than in IEC 61508-4. The definitions are intended to drop in directly for the term, and defined terms can be used in other definitions. In reading the ISO 26262 standard, the terms are used as they are defined, so common usage outside the standard is not enough to understand a term's meaning. For example, *item* is defined as a system or system of systems at the vehicle level. This is not the everyday use of the term, but it is critical to understanding the standard. Hazards can only occur at the vehicle level, so a hazard and risk assessment may only be performed on an *item*. This use of definitions specific to ISO 26262 context affects other terms as well. *Item* is a key term.

Automotive System Safety: Critical Considerations for Engineering and Effective Management,
First Edition. Joseph D. Miller.

Automotive safety practitioners know this and therefore use the terms carefully. When this author taught an introductory course about ISO 26262, he used the terms exclusively as they were defined in ISO 26262, just as if he were teaching a foreign language. Definitions are key to understanding the standard. Still, there may be ambiguity between the reader's understanding and intended meaning of the standard. Some key terms follow.

Architecture

When discussing *architecture,* it is important to note that assignment of requirements to architectural blocks is included. This may be different from a designer's common understanding, so he may be surprised when his architecture document is assessed for compliance with ISO 26262. The *architecture* can be of an *item* or an *element*, and the term *element* is used in the definition. So, what is an *element*? An *element* is a *system*, *component, hardware part*, or *software unit*. Therefore, essentially, an *element* is anything that's not an *item*. This is the way *element* is used in the standard, and it is clearly ambiguous.

ASIL

The *automotive safety integrity level (ASIL)* is one of four levels assigned to a requirement, with D representing the most stringent and A the least stringent level. This term is commonly misused because the *ASIL* is assigned to the requirement, not the *item* or *element*. There is no such thing as an ASIL D microprocessor. Because of this common misuse, the term *ASIL capability* was added that can be used with an *item* or *element* that is capable of meeting ASIL requirements. For example, a microprocessor may have *ASIL D capability*. This is useful for a *safety element out of context (SEooC)*. The safety manual may specify *ASIL capability*, and conditions may apply.

ASIL decomposition is defined as decomposition of requirements with respect to an ASIL, not decomposition of the system architecture, as it is sometimes misstated. Since, as stated previously, the architecture includes the assignment of requirements, the architecture is affected because of the required independence to enable ASIL decomposition. Also, ASIL decomposition does not affect hardware metrics. Automotive safety specialists know this well and take it into account when supporting architecture development. This enables, for example, having QM(D) software performance requirements and D(D) safety mechanisms. Assessment still requires ASIL D independence, and there are other exclusions as well.

T&B Terms and Other Terms

Several definitions were added to ISO 26262 in the second edition for use by the truck and bus (T&B) industry. One of these is *base vehicle*, which is not the same as an automobile without optional equipment. It is the T&B original equipment manufacturer (OEM) configuration before the *body builder* takes it and adds *body builder equipment*, such as bus bodies or a truck cab, to make a *complete vehicle*. This is then the object of *rebuilding* for a different purpose. *Rebuilding* includes a *modification* of a *T&B configuration*. *Power take-off* is defined as the interface where a *truck* or *tractor* can supply power to equipment and is the boundary of the scope for ISO 26262 for T&B. After a period of service, the *T&B configu-*

ration may undergo *remanufacturing* to the original specification. In use, there is a *variance in T&B vehicle operation* due to load or towing, etc. *Truck, tractor*, and *bus* are also defined. Thus, a T&B syntax emerges.

Baseline is defined as a set of work products before change. This is different from a baseline product configuration but is necessary to establish a *safety case* for future use. ISO 26262 defines *candidate* as an *item* or *element* that is identical or nearly identical to one already in production and is considered for proven-in-use credit.

ISO 26262 defines *confirmation measure* as a *confirmation review, assessment,* or *audit* concerning *functional safety*. Thus, these confirmation measures are not the same as technical reviews and audits, as they relate to compliance and not performance. Therefore, a technical review should precede a *confirmation measure*. For a given work product, the functional manager normally approves before the safety assessor.

In ISO 26262, *controllability* is defined as the ability to avoid the specified *harm*, which is damage to people. Thus, avoiding injury relates to *controllability*, not avoiding an accident. For example, due to the differential speed in a rear-end collision, the controllability for ASIL C safety goals are different than for ASIL A safety goals. This is different from the meaning in other contexts, and discussion was required in order for the committee to agree.

Coupling factors is defined in ISO 26262 to support an extensive discussion of *dependent failures* included in the standard. These are common characteristics in *elements* that cause a dependence in their *failures*. This dependence can be triggered by a *dependent-failure initiator* (DFI). *Dedicated measure* is defined in ISO 26262 as something done to ensure that a failure rate is not compromised: for example, by a bad lot of parts. Burn-in is a *dedicated measure*. A *dedicated measure* is not a built-in safety mechanism; it is intended to be used, for example, when the built-in safety mechanisms are not good enough. Then *dedicated measures* are specified.

FTTI

A *fault-tolerant time interval (FTTI)* is defined in ISO 26262 as the time between the occurrence of a fault anywhere in an item that can lead to a potential hazardous event and the occurrence of that potential hazardous event. After substantial debate in the ISO 26262 committee, this was agreed on because hazards can only occur at the item level. Therefore, an *FTTI* at a component level, as *FTTI* is commonly used in practice, is inconsistent with this definition because hazards do not exist at the component level. To support the *FTTI* definition, other time intervals are defined in ISO 26262. First is the *diagnostic test interval*, which is the time between diagnostic tests, including the test time. Then is the *fault detection time*, which is the time to detect a fault, which is one or more than one *diagnostic test intervals*. Next is the *fault reaction time interval*, which is the time between detection of a *fault* and reaching a *safe state* or *emergency operation*. The *fault handling time interval* is the sum of the *fault detection time interval* and the *fault reaction time* interval. The *emergency operation time interval* is the time between the *fault handling time interval* without unreasonable risk, perhaps due to its limited duration that is limited by *emergency operation tolerance time interval (EOTTI)*, and the beginning of the *safe state*. The intent is that the *fault handling time interval* should not exceed the *FTTI*. This syntax is expected to add clarity, and broad implementation is foreseen.

The *fault, error, failure* taxonomy is commonly understood in automotive practice. Every *fault* does not lead to an *error*, and every *error* does not lead to a *failure* to meet a safety requirement. It is expected that *errors* are what safety mechanisms detect. Therefore, the FTTI syntax just described was agreed on.

Motorcycles

Motorcycle is defined in ISO 26262 as a two- or three-wheeled motor-driven vehicle weighing less than 800 kg that is not a moped. To support *motorcycles* being included in the standard, a *motorcycle safety integrity level (MSIL)* was defined as one of four levels to specify an item's or element's necessary risk-reduction requirements that can be converted to an *ASIL* to determine *safety measures*. To determine *controllability*, an *expert driver* is defined as a person capable of evaluating *controllability* on a *motorcycle*. The expert rider is expected to have special handling skills and knowledge. This person can evaluate characteristics considering a representative rider, and then controllability is determined.

Semiconductors

Several of the terms in ISO 26262 are introduced to support guidance for semiconductor components. A *hardware subpart* is defined as a logical division of a *hardware part*, like an arithmetic logic unit in a microprocessor. A *hardware elementary subpart* is defined as the smallest portion of a *hardware subpart* that is used in safety analysis, such as a register. *Observation points* are defined as outputs to observe potential *faults*. *Diagnostic points* are defined as the signals that can be observed concerning the detection and correction of *faults*. *Processing element* is defined as a *hardware part* for data processing, like a core. *Programmable logic device (PLD)* is defined. *Transient fault* is defined like a single bit upset. These definitions support semiconductor guidance and are commonly understood.

The remainder of the terms defined in ISO 26262-1 are primarily to support safety discussions in the standard. *Safety* is defined as the absence of *unreasonable risk*. *Functional safety* is *safety* from *hazards* caused by *malfunctioning behavior*, which is unintended behavior of an *item* with respect to its design intent. A *safety goal* is a top-level safety requirement based on the potential *hazards* from the *hazard analysis and risk assessment (HARA)*. The *safe state* is defined as the *operating mode* in case of *failure* that is free from *unreasonable risk*. Therefore, normal operation is not referred to as *safe state* unless the *item or element* fails operational without *degradation*. This definition was agreed on by the ISO 26262 committee to support fail-operational systems anticipated to be in demand for higher levels of automated vehicles.

ISO 26262-1 defines *functional concept* as the specification of the intended functions and their interactions to achieve the desired behavior. The *functional safety concept* is defined as the specification of the safety requirements, their assignment to *elements* in the *architecture*, and their interactions to achieve the safety goals. The idea in the early days of ISO 26262 development was for the functional safety concept to be provided by the OEM, and the technical safety concept to be provided by the supplier. The *technical safety concept* is defined as the specification of the *technical safety requirements* and their allocation to *system elements* with associated information providing a rationale for functional *safety* at the *system* level. So, the *technical safety concept* was anticipated to be part of the quote from each supplier, and the OEM could then choose. In practice, sometimes the *functional safety concept* is provided by the supplier and sometimes by the OEM, and sometimes the OEM mixes some *technical safety requirements* with the *functional safety requirements*. The supplier provides the *technical safety concept,* but all the *technical requirements* are provided after award of the business, particularly in a joint development.

Terms are defined in ISO 26262-1 to support discussion of the safety lifecycle of software. The first is *branch coverage,* defined as the percentage of branches of a control flow executed during a test. *Modified condition/decision coverage (MC/DC)* is defined as percentage of all single-condition outcomes that independently affect a decision outcome that have been exercised in the control flow, and a note relates this to testing. *Calibration data* is only defined for use in software: it is the data that will be applied as software parameter values after the software build in the development process. Co*nfiguration data* is defined as the data that controls the element build; a note to this entry says it controls the software build. An entire normative annex discusses the safety of using *calibration data* and *configuration* data in manufacturing. *Embedded software* is defined as fully integrated software to be executed on a *processing element*. *Formal notation* is defined as a description technique that has both its syntax and semantics completely defined. *Informal notation* is defined as a description technique that does not have its syntax completely defined. *Model-based development* is defined as development that uses models to define the behavior of the *elements,* and a note to this entry says it can be used to generate code. There is an informative annex devoted to *model-based development*, for software development.

Safety Manager

ISO 2626-1 defines *safety manager* as the person or organization responsible for overseeing and ensuring the execution of activities necessary to achieve *functional safety*. The *safety manager* may or may not manage people and may perform many of the activities to achieve *functional safety*. In many organizations, the *safety manager* is not involved in SOTIF, while in other organizations SOTIF is added to this person's responsibilities. Identification of a *safety manager* from each party is necessary for the execution of a *development interface agreement (DIA),* defined as an agreement between customer and supplier in which the responsibilities for activities to be performed, evidence to be reviewed, and *work products* to be exchanged by each party related to the development of *items* or *elements* are specified. These *work products* are compiled and used as evidence together with an argument for safety to produce the *safety case*. The *safety case* is expected to result from the execution of the *safety plan* that is used by the *safety manager*. ISO 26262-2 is all about planning.

Other definitions in ISO 2626-1 also support the standard, but are perhaps not quite as specialized as the definitions discussed

ISO 26262-2 – Management of Functional Safety

Safety Culture

ISO 26262 starts out discussing the requirements for each organization to have overall rules and processes for functional safety in addition to a safety plan for a specific project. In automotive practice, these overall rules are often audited prior to a business award for a safety-related automotive product, like a steering system, braking system, or driver assistance system. The organization is expected to institute and support a safety culture and have good communication; an annex is provided to give hints of what to look for in a good safety culture, and tips on how to spot a poor safety culture. The organization is to have a good process for resolving product issues, have good training, and have a good-quality organization. While this all relates to safety, it also relates to deciding about new business awards. These organizational attributes are normal automotive practice and are necessary for business.

Some of the indicators of a poor safety culture include not having traceable accountability, not prioritizing safety and quality over cost and schedule, undue influence of management over safety and quality assessors, overdependence on testing at the end rather than executing safety activities on schedule and allocating resources when needed, management only reacting to field problems, ostracizing dissenters, and ad hoc processes. The keys to a good safety culture address these and are expected to be confirmed in auditable documents in practice. The standard requires evidence that each of these areas is addressed by the organization: specifically, compliance with ISO 26262 is mandated by the organizational rules, and the resources are available. The expertise must be available for assessing safety tasks. In practice, some organizations use internal assessors and auditors, some use external assessors and auditors, and some use both. ISO 26262 only requires skill and independence from the release authority; there are numerous ways to comply.

Cybersecurity

Concerning communication, it is required that cybersecurity and safety communicate in order to ensure that any hazards that can be maliciously induced are identified and mitigated. Organizationally, this is challenging as the domains are separate and can execute their tasks independently. Sometimes, cybersecurity organizational management is connected to the safety organization and can be actively managed. Communication can be mandated in internal processes. An auditor periodically facilitates communication, because failure to communicate is possible.

Quality, Management, and Certification

Evidence of a quality management system is auditable for safety. In practice, it is practically impossible to be competitive in the automotive business without an excellent quality system. There also must be evidence of competence management. In practice, this is usually established by means of training, either internally or externally, of personnel assigned to product development, as well as management, purchasing, and personnel. Personnel keep the records. Certification of key personnel is also used quite often as evidence of competence but is not required by ISO 26262, and ISO 26262 does not provide guidance for certification bodies. For over 12 years, the founder and leader of the US delegation for ISO 26262, who was also a project leader for part 1, was not certified. Nevertheless, there are reputable certification bodies that provide training and require demonstration of competence before awarding certification. There also must be evidence of a "safety anomaly" reporting system. In practice, this is included in a problem or project issue management system for the organization, with specific language related to safety added. This can also be supplemented in the overall organizational safety document that is required. A reference that is followed is needed, and internal auditing supports compliance with it.

In addition to an organizational safety plan, ISO 26262-2 requires safety management at the project level. This is an instantiation of the overall organizational safety plan or a separate process followed by the project. In practice, for an enterprise that has multiple product lines and many projects, some perhaps out of the scope of ISO 26262, a more generalized organizational plan may be developed. Those product lines that are in scope of ISO 26262 share a common plan tailored for each project. The safety plan is tailored for the infrastructure of each product line, if applicable, to improve efficiency. Each project safety plan must meet the goals of ISO 26262-2. Roles and responsibilities must be clearly assigned: a generalized plan with no names, names missing, or disputed responsibilities is unacceptable at the project level and should be escalated to executive management when audited. The plan needs to include an impact analysis of the item or element, to determine whether the item or element is new or has some content carried over from previous projects. In practice, nearly all automotive projects contain some "carry over," even "new" projects, and the carried over content has some modifications or changed environment. These modifications need to be addressed in the impact analysis to determine whether the changes are safety-related and what tailoring to the safety lifecycle is to be included in the safety plan. The safety plan must support tracking of progress, distributed developments, and creation of a safety case, including the arguments for achieving safety, an assessment of safety, and a decision as to whether the safety achieved is enough to justify series production and public sale. In practice, this requires consistency and review of the impact analysis and plan. For example, every project may be a carryover; only the microprocessor is new, the software is modified to include a new feature and interface, and manufacturing is at a new site with advanced processes. Every lifecycle phase is impacted, but it is still it is a carryover.

In order to produce a safety plan for a project, organizational rules must first exist, competent resources must exist, a project manager must exist, and the project manager must appoint a safety manager. If the project is a distributed development, the customer and supplier must both have safety managers. The safety manager (who is the project manager unless another safety manager is appointed) must make the safety plan with the safety activities, including assessment of work products with independence as required in ISO 26262-2, auditing, and a final assessment and signoff for production. Operation safety activities are also included. In practice, this safety plan content is often included in a template or tool and tailored for the project. The supplier may assign some of the work product content from ISO 26262 to work products that exist in the enterprise development process. The customer may do the same thing. Then the safety managers of the customer and supplier work out an efficient way to satisfy both their processes so as not to increase cost. Some work products are the same, and others combine content differently.

Impact Analysis and Safety Plan

Every project within the scope of ISO 26262 requires that an impact analysis be performed at the beginning and that it have independent assessment at the highest level of independence (I3), regardless of ASIL. This requires an item definition to define the scope. In practice, the person who performs the impact analysis needs significant domain knowledge of the product and significant safety competence. Often the safety manager creates the impact analysis. Even if parts are carried over, there needs to be enough documentation from the reused item for the safety case unless it was developed, or was in development, on or before November 15, 2011, when ISO 26262 was first released. Thus, conditions for documentation that may have been allowed in the original release are expected to be corrected in subsequent releases of the item. The item may have proven-in-use content that allows tailoring, though that is somewhat rare in practice. More common is the use of an SEooC, which allows lifecycle tailoring if the SEooC assumptions of use are met.

When tailoring of the safety lifecycle is justified based on the impact analysis, then it is done work product by work product, with a rationale for each. In practice, the impact analysis references the work products used by the enterprise rather than the standard, and the tailoring is directly applied. The enterprise usually has fewer work products than the standard, so fewer justifications are needed. The impact analysis is never tailored: it justifies the plan.

The plan is tracked and updated at least at the beginning of each phase. In practice, it is updated at each incremental audit and perhaps also at each reporting period if the enterprise agglomerates the safety-case status for executive review. This works because the safety case is incrementally built using the work products and arguments in each phase. Problems are spotted early, which helps achieve product launches.

For the safety case to be on time, the confirmation measures need to be passed on time. The practical hurdle here is getting the key work products scheduled in time with an independent assessor not under the project manager's control and allowing enough time for the assessment and any rework that is required. Assistants can help get the confirmation reviews done efficiently, but the lead assessor always makes the final judgment. Techniques also exist to streamline efficient audits and final safety-case review and approval. In practice, incremental audits focus on issues related to complying with the organizational safety plan or policy, and executing the process, as well as follow-up on corrective actions and recovery plans. This incremental approach solves problems in practice. Confrontational rejections at launch are usually avoided, but they are still possible.

The safety plan also contains production-related work products such as evidence of a quality management system. In practice, this is often achieved by reference to the quality system certification, achieved by most automotive suppliers, for the facility or facilities involved in producing the item or element. There needs to be a person responsible for functional safety in operations. In practice, this is the quality manager or engineer, or the customer or supplier liaison at the facility. This person requires safety training as well as product safety training and training concerning the legal aspects of product safety in the nation where production is located and the product is sold, though the latter is not specifically required by ISO 26262. Meeting safety requirements for service and decommissioning must also be planned. In practice, this is accomplished with warnings and instructions for the user and service manual that are normal practice for automotive. Customers expect supplier support, and vehicle manufacturers (VMs) normally provide it.

ISO 26262-3 – Concept Phase

Initiation

The concept phase in ISO 26262-3 starts with the item definition. This is a system description that is used in further safety analysis on the project. In practice, this is usually created by system engineering and, for a supplier, is expected to be consistent with the information system engineering communicated to the customer, including its expected performance at the vehicle level. In addition, any applicable regulations or standards, known hazards, systems with which the item interfaces and assumptions about them and vice versa, distribution of functions, required quality, performance including capability of any actuators, important operating scenarios, and availability are included in the item definition. It is important to get the item definition done right away, because the impact analysis and all the other work products depend on it, either directly or indirectly via risk of inconsistency. Practically, this is contentious because of the prioritization of system resource allocation early in the project. Even a draft may prevent project lateness and can be updated as needed.

It is also necessary to perform the hazard and risk analysis early in the concept phase in order to identify the safety goals and their ASILs. The item definition is necessary to get the HARA started. In practice, much of the HARA is carried over from previous items that are being modified; the HARA is modified for the changes, such as an additional interface between the steering system and the headlights, or a new fail-operational requirement on the steering system to support an automated driving feature. The safety goals are needed to enable flow-down to any further requirements, or new ASILs for old requirements, that may affect the basic concept of the item and its functional architecture.

Sometimes the customer will perform the HARA; the supplier can also perform the HARA or may have a HARA they plan to reuse. When this happens, there is a high likelihood that the separate HARAs will not exactly match the first time. This must be resolved to prevent conflicts in the safety cases of both parties. Neither party may want to change, and both may feel they must be diligent to evaluate the reasoning of the other party. Key to this is to consider the operational situations each considered and to agree on the ranking of severity, exposure, and controllability. Actual tests and traffic statistics may help, but often legacy ASILs prevail, as in the annexes of SAE J2980. For new features, systems, or functions, a hazard and operability study (HAZOP) approach, as described in SAE J2980, considering traffic statistics, provides the most thorough and defendable HARA. If the bias of the analyst is to assign a severity higher than the traffic statistics justify (e.g. S3 instead of S2 if a random accident from a loss of control results in death or life-threatening injuries 7% of the time), then the exposure might reflect exposure to situations that result in this harm (e.g. E3 instead of E4), consistent with SAE J2980 guidance. Hazards outside the scope of ISO 26262 are considered, but failure of external measures is not considered. When setting up the failure modes and effects analysis (FMEA) for top-level system-failure effects, the hazards identified in the HARA are included. There usually are other system failure effects, and the FMEAs should be consistent.

Reasonably foreseeable misuse is an important consideration in the concept phase. This is different, but difficult to distinguish, from abuse. For example, it is foreseeable that drivers may take their hands off the wheel in a driver assistance system in situations where the system is intended to maintain the lane and following distance. A malfunction may result in a random collision due to a loss of control, due to this misuse. A mitigation is to monitor torque on the handwheel, warn for a period, and then disable the system with adequate warning. This reduces exposure to a reasonable level. Abuse is to tape a beer can to the wheel to deliberately defeat this failsafe mitigation. Such behavior is classified as abuse rather than reasonably foreseeable misuse, and it is not considered further in the concept phase. Automotive practice is to verify the HARA and safety goals with the design team, have an independent confirmation, and verify with the customer. Safety goals may combine hazards or be very granular; it is a trade-off with complexity. Either approach can work.

Functional Safety Concept

After the HARA has identified the safety goals and ASILs, the customer, or the supplier for an SEooC, can start the functional safety concept to determine how to meet the safety goals. The VM often, but not always, develops a functional safety concept for an item because the VM can assign requirements to the various items in a way that takes advantage of their independence. In practice, this is often done in collaboration with the suppliers – for example, in a joint development. Past experience often allows much of the functional safety concept to be carried over, with updates for new features and functions. In the end, the functions, degraded behavior, detection and control of faults to satisfy the safety goals, allocating requirements to the functional design, and validation criteria must be determined. In practice, due to experience or cost considerations, some detailed requirements are also included that are expected to be technical requirements of the design instead of conceptual requirements; these then become design constraints. All safety goals must be addressed by the functional safety concept. The requirements are assigned to architectural elements, using architectural assumptions.

The functional safety concept must, at minimum, if applicable, have functional requirements that specify fault avoidance, fault detection and control or the control of the resulting malfunctioning behavior, transitioning safely to a safe state, including any emergency operation, fault tolerance, degraded functionality and warnings, the FTTI and related detection and fault-handling intervals, and arbitration of multiple requests. All the operating modes and redundancies need to be considered. Therefore, even a relatively simple functional safety concept can be complex and warrant collaboration. In practice, the functional safety concept is even more complex and is extended to include the technical safety concept with requirements that govern a specific application. Any actions expected by the driver are documented in the functional safety concept. Requirements assigned to other technologies, such as the availability of a manual steering system if the electric steering assist fails, are specified in the functional safety concept. In practice, if any of these are missing from the VM's functional safety concept, the supplier will probably add them anyway. The VM may add many more requirements, some quite detailed. The functional safety concept from one VM rarely matches that from another VM for essentially the same system. The supplier needs to manage this complexity and provide safe systems for all the VMs. Thus, the supplier may have a generic functional safety concept for reference, and manage the differences. This approach is less complex.

The functional safety concept needs to be verified and validated. To verify the concept, often hazard testing is performed to specify the hazard in engineering terms. For example, a prototype system, such as a steering system with test software, can be installed in an equivalent vehicle and tested to establish the FTTI. This may be a test of how long a period of unwanted full-assist pulse must be present to later cause the vehicle to cross the lane boundary from a specified lane position. The pulse is applied and removed in gradually increasing increments per test, and then the vehicle position is monitored after each pulse for several seconds while the driver keeps their hands on the wheel. Then a mitigation is checked to see whether the hazard is avoided when that FTTI is achieved by the concept vehicle. If it isn't, then improvements are made to the concept and prototype, such as shortening the detection and reaction times, and the tests are repeated. This hazard testing is best done early in the concept phase, because the design implications can be significant. In practice, the supplier does this on a representative supplier vehicle. The hazard testing gets updated for the application, and the customer facilitates this.

ISO 26262-4 – Product Development at the System Level

Technical Safety Concept

After the functional safety concept exists, the technical safety concept can be developed. Even if a supplier has significant experience with the item being developed, the technical safety concept cannot be finished, without risk of error, until the impact of the vehicle architecture and feature content has been resolved in the functional safety concept. Then the design architecture can be confirmed, and the technical requirements can be elicited and assigned to the architecture. In practice, there are three sources of requirements for the technical safety concept: the functional safety requirements, requirements elicited from design analyses, and requirements elicited from the assumptions of others, such as the mitigations assumed in the safety manual of the microprocessor. The hallmark of the technical safety concept is that it breaks down and assigns requirements to hardware and software that fulfill the functional safety concept. In practice, this is nearly always reviewed by the customer's safety team. It is an early deliverable, and the customer wants it quickly.

The technical safety concept must contain the requirements for all the safety mechanisms needed to prevent failures due to single-point faults, latent faults, and dependent failures. These are normally found through the single-point and latent-point failure analyses. However, at the time the technical safety concept is required, these analyses have not been performed. If the item is a modification, the old single-point fault metric (SPFM) and latent fault metric (LFM) analyses are a reference, and adjustments can be made based on the modification. Otherwise, the technical safety concept is analyzed qualitatively and then updated after the SPFM and LFM analyses are complete. The technical safety concept may also specify measures to prevent dependent failures or independence between elements to allow decomposition of requirements with respect to ASIL. Requirements on other systems can be specified, but this normally requires an update of the functional safety concept. In practice, any requirements assigned to other items require agreement with the VM. An example is for interfaces with other items to include a range check to prevent propagation of an error that could result in a failure, like a steering system range-checking a lane-keeping assistance request from a camera. Such checks are normally independent, and ASIL decomposition is supported.

When the diagnostics necessary for meeting the SPFM and LFM are included, then the details of the hardware-software interface must also be included, as well as the operating modes. Shared and exclusive resource usage is specified. All the top-level requirements are necessary that allow the hardware and software to be designed. Also, the requirements for production, operation, maintenance, and disposal need to be specified when the architecture is specified. These requirements include special characteristics to be controlled, requirements for diagnostics to provide data for the maintenance personnel, warnings and instructions to ensure safe maintenance and restoration of the item to safe operation, and requirements for the design and instructions for decommissioning, such as the interface to a special tool to deploy the airbags during disposal. In practice, the technical safety concept and the requirements for production operation, maintenance, and disposal are updated in other safety lifecycle phases. They can be created in a requirements tool: all these requirements need to be verified, and the tool facilitates traceability.

Integration, Verification, and Validation

Another phase at the system level is integration, which requires an integration strategy and criteria for acceptance. In practice, this is done in three steps, much like the description in the standard. After the software is integrated together, the integrated software is integrated to the hardware. All the hardware-software interfaces are verified. This is usually when all the safety mechanisms are verified: for example, by fault insertion. The startup and shutdown safety mechanisms are also identified. Communication interfaces can be simulated and verified, including corrupted messages and stuck messages. Any other software safety requirements are also verified. The methods for verification are specified in the strategy so they are compliant with ISO 26262. These compliant verification methods can also be captured in the requirements tool or be a separate document that is referenced. When all this is checked out, other systems can be interfaced if they are needed to create the item. Intersystem diagnostics and safe states can be verified, with different representative calibrations included as required. When the item is finally working and verified, it can be integrated at the vehicle level with other items: this is accomplished in much the same way as the previous integration, but usually with the VM. Different operating modes, normal vehicle behavior, and extreme behavior are verified. This verification may confirm that false detections are avoided. Any improvements are made.

Validation is normally executed on the vehicle using a validation procedure, which must include a way to validate that each safety goal is achieved. In practice, this is usually executed using the hazards from the HARA and performed with the customer. Faults are inserted under different conditions, and controllability is evaluated. For example, a fault can be inserted during a steering maneuver that results in a loss of assist. The driver controls the vehicle and stays in the lane. The test procedures and pass/fail criteria must be determined in advance, and the results are evaluated using those criteria.

ISO 26262-5 – Product Development at the Hardware Level

Requirements and Architecture

The hardware development phase of the safety lifecycle details the hardware implementation of the technical safety concept, analyzes the hardware design for potential faults and their effects, and coordinates with the software development. The hardware phase starts with detailing the hardware requirements from the safety concept. In practice, most automotive hardware is a modification of previous automotive hardware. The effort is shortened to determine what requirements have changed in the technical safety concept, what changes are planned in the hardware for other reasons, and then, from the impact analysis, what needs to be updated in the hardware lifecycle, including work products. If new functions require new hardware, the technical safety requirements for the hardware related to that feature are flowed down to hardware requirements that can be used for design and verified. For example, if the hardware has a new microprocessor, then any safety assumptions, such as an external watchdog, need to be specified. In a new design, all the detailed hardware requirements are elicited. The hardware-software interface is also specified and communicated to software development, and will be verified as well.

The hardware design starts with the hardware architectural design. All the hardware requirements are assigned to architectural elements; the highest ASIL of the requirements assigned to an element determines the ASIL assigned to all the requirements of an element, unless the requirements for independence are fulfilled. In practice, this is especially relevant for custom integrated circuits that have several safety-related requirements of different ASILs as well as some non-safety-related requirements. The highest ASIL is applied to all the requirements. This may impact verification methods that must be used if independence is not established by means of a dependent-failure analysis (DFA) for cascading failures. Most hardware components do not have a requirement with an ASIL assigned to them. The architecture considers the targets for the SPFM and LFM, and the requirements for the probability of a random hardware failure or the failure-rate class of the components. In practice, these metrics can play a major role in the criteria used to select components for the hardware design. The hardware architecture design considers systematic causes of failure, like temperature, vibration, and ingress of water or other fluids, if relevant. Specific tests can be run before verification to evaluate these systematic causes, like ingress. The environment can influence the tests: for example, if an enclosure may leak saltwater, multiple short circuits are possible, and special measures are needed.

Safety analyses are expected on the hardware design. In practice, an FMEA has always been performed on the hardware in automotive applications, long before ISO 26262. Sometimes the FMEA is a true design FMEA: for example, with preventative measures being the specification and the detection methods being analyses and testing. Other times, the FMEA is a field FMEA, with detection methods being diagnostics; or the field and design FMEA can be mixed. Increasingly, the hardware FMEA is becoming a true design FMEA to detect systematic faults due to the design process; and the field FMEA is replaced by the SPFM/LFM analyses since the diagnostics are redundant, and the SPFM/LFM analyses provides more value. Inductive analysis is also performed: this is a qualitative fault tree analysis (FTA). It could be quantitative.

After the safety mechanisms are identified from the analyses, it must be shown that they work fast enough, that the fault-handling time interval is met, or that the FTTI is not exceeded. In practice, sometimes this is shown in the concept phase during hazard testing. Instead of running many tests to determine what the FTTI is, in a product with carryover safety mechanisms, a fault is inserted, and it is determined whether the potential hazard is avoided at the vehicle level. Then, during design, the timing of the rest of the safety mechanisms is re-verified. The needed evidence of safety is thus created. This process can be more efficient than repeating hazard tests to determine the FTTI.

All the hardware safety requirements need to be verified. In practice, the means to do this is usually captured in a requirements tool and linked to the requirement in order to show traceability. This is needed not only for the safety mechanisms, but also for the hardware-software interface, including the scaling that allows in-range and out-of-range signals to be evaluated, as well as the assumptions of any SEooC included in the design. Simulations are often used to support design reviews: they allow tolerance analysis to be reviewed, and the evidence can support the safety case. The safety case requires retrievable, referenceable documentation.

As the analyses are performed on the hardware design, it is automotive practice to determine whether the design characteristic being reviewed is critical for production. In practice, many critical characteristics are already included in the drawings, such as alignment of a sensor. In addition, as the hardware FMEA is performed, criteria can be established for characteristics to become candidates to be special or critical characteristics. If they are, then production criteria must be established. In practice, this is done at an enterprise or product-line level in order to have consistent judgment and processes. Often the processes are statistically controlled, with more stringent statistical criteria for special or critical characteristics. Field traceability for hardware parts or lots is common automotive practice; such automotive practices comply with ISO 26262-5 and are long established.

SPFM and LFM

Hardware architectural metrics specific for automotive practice were introduced by ISO 26262. Now they have become common practice and are nearly always reviewed in audits. They are a critical consideration when evaluating an SEooC for use in an automotive project. Even for automotive practitioners that previously performed an SFF per IEC 61508, the SPFM/LFM has differences. For example, a two-channel redundant architecture can be decomposed into two SIL 2 channels, each of these channels with a 90% SFF for a SIL 4 system with a fault tolerance of 1. The same architecture, referring to ISO 26262, would need to show 99% SPFM considering both channels together and have an LFM of 90%. In practice, significant effort is devoted to hardware architectural metrics. Often, to reduce the effort, violating more than one safety goal at a time is considered when performing the analysis. This allows reuse of the SPFM and LFM for more than one customer, since often the safety goals of different customers are expressed differently for the same system; the result is a more conservative analysis. Analyses tools can also improve efficiency and have built-in reliability references. Sometimes field data is used, especially from IC suppliers, since it is more accurate than the references; this is especially true of transient fault rates. The targets for the SPFM and LFM can come from the table in ISO 26262-5 or from a similar trusted item in the field. To determine targets based on fielded systems would require that an SPFM/LFM analysis be available or performed on the item in the field. So in practice, the table is almost always used to set the targets.

Requirements for safety mechanisms are elicited, and communication and traceability of the requirements follows; many are software requirements.

Random Hardware Failures

In addition to the SPFM and LFM, there must also be an evaluation of safety goal violations due to random hardware failures. While the method that may come to mind to perform this evaluation is a quantified fault tree of the hardware, efficiency is gained by computing the SPFM, LFM, failure rate classes, and probability of failure due to a random hardware failure in the safe tool, such as a spreadsheet or commercial tool. Then each component is evaluated only once: the safety mechanisms can be determined, taking all the metric requirements into consideration. Criteria are set up to highlight potential issues requiring further consideration. The targets for the probabilistic metric for random hardware failures (PMHFs) are determined from a trusted design in the field or the table; in practice, they are usually selected from the table. In addition to the PMHF, the *cut set method* or *second method* that considers diagnostic coverage and residual dangerous failure rate of each individual component has gained preference by some supplier analysts and is becoming generally acceptable to customers that traditionally relied on the PMHF. As discussed in Chapter 9, recalls are not often caused by "random" hardware failures. Suppliers may prefer the second method because it helps isolate and resolve potential issues with a single component and establish measures to prevent a "bad lot" of parts from causing field problems. Also, the second method is easier to maintain than the PMHF since only parts affected by a change are reanalyzed. If several systems are combined to make an item, the PMHF requires allocation of the target, while the second method does not. Nevertheless, many practitioners use the PMHF, and this legacy may linger.

Hardware Integration and Verification

Hardware integration and verification has been an important part of automotive practice since long before the release of ISO 26262. The results are always reviewed by the customer. There is substantial reuse of hardware circuits, either informally or by means of a formal process. Therefore, new and reused hardware is integrated before the assembly is tested; this is performed hierarchically if appropriate. It has been automotive practice to perform functional testing and environmental testing during hardware verification. Now testing has become traceable to all requirements, and test cases are assigned to verify all safety requirements. Often this is audited prior to the start of design verification, including fault-insertion testing. Other test cases are assigned in accordance with the standard and past field experience. The verification evidence of each SEooC is taken into account. Corrective actions may result in modifications and retesting, and require recovery planning.

ISO 26262-6 – Product Development at the Software Level

Development of automotive software can be the most labor-intensive part of an automotive project. There are massive amounts of safety-critical software on vehicles. Development of this software requires development of the architecture and coordinating the efforts of system design, hardware design, safety specialists, cybersecurity specialists, and many software design engineers. Compatible tools, including system and software modeling tools, must be used across all disciplines. The software needs to be designed to consistent guidelines and standards, including style guides, naming conventions, and consistently used defensive programming techniques. Concurrency of processes and tasks requires attention, and the interfaces need to be understood.

Software Development Process

Software development generally follows a classic V, but agile development processes are also employed. Still, software development is usually independently audited to the automotive SPICE standard, and achieving the required maturity level is a condition to be met by the supplier in order to receive future business awards. The software schedule, including verification, is planned based on a different schedule than hardware and the rest of the project; coordination with hardware development and the delivery schedule are planned into the software development schedule. This software plan can be used to plan compliance with ISO 26262-6. The work products will be confirmed, and the process will be audited.

Automotive software development, as expected by the deployment and rigorous enforcement of the automotive SPICE maturity model, is driven by requirements. Software safety requirements are elicited and refined from the technical safety concept. The software requirements to meet product functionalities with the needed properties so that the safety requirements can be met are also specified. This includes, for example, sensor scaling and control scaling so that malfunctions are detected by the safety mechanisms. The software requirements are specified to comply with the requirements of the hardware-software interface. Each of these requirements must have a verification method and criteria assigned to it and must also specify the architecture requirements, including any independence requirements needed for decomposition of requirements with respect to ASIL. A DFA is required to support ASIL decomposition. Often the hardware provides support for independence; this is included in the DFA.

Software requirements need to contain requirements elicited by software safety analysis, assumptions derived from hardware safety analyses and assigned to hardware, and the requirements derived from the assumptions of any SEooC used in the design. For example, requirements for startup and shutdown diagnostics are specified to control latent faults. Usually a tool is used to manage software requirements that can link directly to the technical safety requirements for software. In automotive practice, these requirements are audited: evidence of completeness is sought, and traceability is checked.

Software Architecture

In practice, the software architecture in automotive is often specified using semi-formal notation. There are tools that support this, though not all tools are seamlessly compatible with each other, so some trade-offs are needed. Advantages such as retaining traceability to the requirements database are considered as well as support for safety analyses and documentation. While the software can be complex, sometimes supporting multiple processors and multiple core pairs, comprehensibility is sought from a hierarchical architecture. This allows review and presentation to the auditors and customer and helps avoid systematic errors. Trust is built when the requirements are easily traceable. The tools and hierarchical architecture help traceability. Demonstrating completeness of requirement traceability requires metrics that may also be tool supported.

There are many measures as to how good the architecture is: complexity metrics may be required by the customer, a scheduling analysis is almost always required, and the architecture is expected to go down to the software units. Traditional good practices, such as size restriction and encapsulation, are expected. Testability of the architecture is a consideration, and there are safety requirements for test coverage in ISO 26262. Also, calibration for different models to be supported in the software application is commonly provided in automotive software. These calibrations must also be verifiable and defendable; otherwise, field problems may arise.

After the architecture is established and the units are defined, then the requirements for the units are implemented in the unit design. In practice, the units are carried over from another design, coded by hand, or implemented with code generated in a model-based design. If the latter is true, then the requirements of ISO 26262 are applied to the model. Otherwise, they are applied to the code. These requirements are commonly accepted.

Verification

Software unit testing of every software unit is common automotive practice, though other verification methods are specified in ISO 26262 and used. Testing of the safety mechanisms and interfaces to hardware and other units are verified. Back-to-back tests are often used because the use of model-based development is becoming common. Often, unit verification is not conducted in the target environment: it can be on a simulator, for example, and differences from the target environment can be justified. Each requirement assigned to the unit is verified at the unit level: test cases are generated and linked to the requirement, pass/fail criteria are set, metrics are determined for MC/DC coverage, and the results are recorded.

Software-to-software integration is common automotive practice. In addition to the software units that have been tested, often software integration testing includes software from other suppliers on the project as well as software from the customer. This is black-box software that has not been unit tested by the integrator, and the integrator then must rely on the tests performed by the other suppliers or the customer. The integration test is performed on the model if all the units and other software to be integrated have been modeled, and then the integrated code is generated after the integration tests. It is still common automotive practice to perform integration tests on the actual code. This integration is often performed on the target processor or on an emulator of the processor to be used. The integration steps are always planned, and the safety mechanisms are always verified. The architecture is verified to meet the functional requirements. Any failures and abnormal, unwanted, or unexpected behaviors are corrected; and issues due to inaccessible software are resolved. Any temporary test software is removed.

The methods used for software integration in automotive practice are generally consistent with the requirements of ISO 26262-6. Software requirements are verified during software integration, and the results are documented and traceable to the requirements verified. This is more practical to do at software integration than at hardware-software integration and provides a baseline for use in troubleshooting any anomalies at hardware-software integration. Resource usage can be confirmed at software integration, as well as timing. Fault insertion is executed, including timing faults if applicable. Back-to-back tests are recommended and are useful in detecting any faults arising from software tools. Control and data flow are also verified. Other techniques can be used for verification.

Integration test cases are derived prior to integration testing, and the pass/fail criteria are specified. Automotive practice is to conduct a safety audit ahead of performing integration testing, and these test cases as well as the availability of pass-fail criteria are checked. Equivalence classes are nearly always used when many calibration combinations are required; they improve the efficiency of testing. Known or past errors that may otherwise be missed, or that are not directly linked to a requirement but may result in unwanted or unexpected behavior in the past, are tested. This unexpected behavior may be as a result of misuse not foreseen in the past, or some other anomaly. Such tests are mandated by a permanent corrective action, even if the original issue was corrected.

In order to show completeness of requirements coverage, metrics are kept on the number of requirements linked and not linked to verification methods, and the number of verification methods that have been passed or failed. This will almost certainly be audited, probably internally as well as by the customer. Automotive safety systems have a very large number of requirements. Often there are duplicate requirements resulting from multiple safety requirement elicitation measures as well as retention of legacy requirements; tooling and systematic reviews may therefore result in a reduction of the number of safety requirements while completeness and test coverage are being determined. Normally this reduction is not a problem in automotive safety audits, but it is an expectation of the auditor. Structural coverage and call coverage are also measured. Complete and verified requirements are the goal – missed requirements can lead to recalls.

Testing the embedded software is considered critically important in automotive practice. It is not unusual for automotive suppliers to make a considerable investment in the test environment to ensure that the embedded software verification provides enough confidence for series production and public sale of safety-related systems, such as braking and stability control systems. Often, such a testing environment supports hardware in the loop tests to ensure that the embedded software performs as expected when integrated with representative off-vehicle system hardware that is used and tooled for executing test cases. This includes representative electronic control unit hardware and a vehicle communication environment. If necessary, a vehicle is used; otherwise it can be simulated.

This testing of the embedded software almost always includes fault-insertion tests, though the standard only highly recommends fault-insertion tests for ASIL D requirements.

Requirements-based tests are required and performed. The test cases are determined in much the same way as the software integration tests were determined. Equivalence classes are used to ensure that all calibration combinations are adequately covered. The different operating modes are verified, including startup and shutdown modes, as well as operation in the specified degraded modes of operation; these degraded modes are triggered by a degraded mode of another system. Startup while other electronic control units are in communication with the software under test is verified, perhaps on a vehicle. Requirements test coverage is ensured and reviewed and is also audited.

Model-based development is broadly used in some form throughout automotive software development. ISO 26262-6 dedicates an annex to explaining the benefits of and cautions for implementing model-based software development. Tools are available to support various levels of model-based development, leading to wider use. Once the software requirements are captured, they are modeled in a way that supports software development. For example, the architecture of the model may mirror the intended architecture of the software, in some cases. The requirements are imported directly into the model from the requirements database. This supports identifying differences in the requirements from an existing model of a previous development. In other cases, the requirements are captured directly in the model, perhaps with utilities for managing the requirements in accordance with ISO 26262.

Once the requirements are modeled, test vectors are generated to obtain the required coverage. These vectors are used to test the embedded code and compared to the result of the model. Since this will detect an error of the intervening tools, such as a compiler, these intervening tools need not be qualified. If the model is executable, it can also be used for code generation. Often, modeling the safety mechanisms and other requirements can be difficult, as suggested by ISO 26262-6. Then the model-generated software can be supplemented by hand-coded software; it is a pragmatic decision and not unusual.

As previously discussed, it is common automotive practice to use configuration data and calibration data in production automotive software. Because of this, ISO 26262-6 dedicates a normative annex to specify the safety requirements for this practice. Configurable software and configuration data are combined in operations to generate the safety-related application software. Then this application software is calibrated for a build of the vehicle: for example, the stability control system software is configured for a model of a vehicle with particular feature content, such as a midsize vehicle with an electronic parking brake, and calibrated for the individual selection of that content, such as the stability control system software being calibrated for a particular engine choice. Errors in any of the steps to realize the software build and calibration can enable a potentially hazardous situation. Safeguards to ensure that the configuration is valid and correct are employed. Likewise, the calibration is verified to be valid, compatible with the configuration, and correct; such plausibility checks are common practice. Data can be stored redundantly and compared.

ISO 26262-6 provides guidance for freedom from interference. This is especially important for automotive software, because decomposition of requirements with respect to ASIL is common practice in software. This practice requires independence assured by DFA. Analysis is recommended at the architectural level to determine whether the architecture could propagate errors, and further DFA is recommended. In automotive practice, such a software system-level analysis is performed quite often. Dependent failures triggered by such faults as livelocks, deadlocks, and the reaction of the software to the exception of a hardware failure are performed. Guidewords are used on each output of the architectural elements, as in a HAZOP. The remedies to potential dependent failures are supported by hardware and software means. For example, consider a fault in read-only memory. Via a software checksum, such an error is detected, and the software that handles such errors and obtains a safe state is triggered. However, if the error is in the software that handles such errors, a common-cause error occurs that leads to a failure. The remedy is to provide redundant software to handle memory errors. Similarly, deadlocks and livelocks can prevent execution of the software. A safe operating system is capable of ensuring timely execution: the routine details are entered into the system. Then this class of failure is removed from the guidewords, and the analysis is more efficient. In addition, hardware provisions can isolate memory access and improve independence.

ISO 26262-7 – Production, Operation, Service, and Decommissioning

Quality Certification

It is common practice for automotive enterprises to be certified as compliant with IATF 16949. This quality certification gives assurance that the enterprise has processes suitable for production of automotive safety-related items that are intended to be installed in automotive vehicles. Certified compliance with IATF 16949 is referenced as sufficient evidence and is often referenced in the safety case documentation as evidence of a quality system. It is also a prerequisite for being considered for many automotive business awards: customers may require quality certification and conduct additional audits.

ISO 26262-7 requires extensive automotive practices to ensure the safety of the product being produced, and the automotive industry has implemented extensive safeguards for this purpose. These measures always include controls of special and critical characteristics included in the drawings and specifications that support the product being produced, including special and critical characteristics of the parts used in the product. In addition, dedicated measures associated with these parts, as identified in compliance with ISO 26262-5, are implemented: for example, by burn-in facilities or by requiring a certificate of compliance with these characteristics from parts suppliers, supported by surveillance audits of these suppliers.

Safety Instructions

Requirements for safety are also included in assembly instructions: for example, sensor-mounting instructions. Handling requirements include safety requirements such as, for example, what to do if an assembly is dropped, or how long parts can be stored. Each configuration of the product has references to the safety requirements, such as which calibrations are allowed for each of the different software builds. If production practices that led to field actions are applicable to the product, then the permanent corrective actions are applied to the current product.

It is automotive practice to provide training for production personnel that includes instructions to implement the safety requirements. All of these steps must be planned before production can begin, including tooling, capability to meet critical characteristics, and training. The plan is implemented before production starts – production planning starts in development.

The planning includes not only process flow and instructions, but also traceability and tooling. Implementation of all dedicated measures is planned. Normal automotive process is to analyze process failures using a process failure modes and effects analysis (PFMEA). It is not uncommon for the safety manager of the product being manufactured to compare the severity of effects in the DFMEA to the severity of effects in the PFMEA for consistency. When high-severity effects are discovered in the PFMEA, mitigations are put into the control plan, such as mistake-proofing as well as more stringent statistical controls that take process centering into account. The process for configuring and loading software always garners much scrutiny and review. Manufacturing engineering usually maintains contact with development engineering and provides feedback to ensure a design that can be produced safely and efficiently. The PFMEA is almost always reviewed with the customer, but it is almost never provided to the customer, due to confidentiality.

In automotive, preproduction samples are often called *C-samples*. These samples are to be off-production tools and processes, and they are often used by the VM for fleet testing. In order to produce such production-representative samples, run-off or run-at-rate parts from the production tooling suppliers can be used. For example, pinions and racks produced at the supplier can be used at the tier 1 supplier before the tools are moved to the electric steering production line. Quality controls similar to in-line controls can be applied as well as assembly controls. The result is production-intent parts: C-samples.

Automotive practice pays special attention to instructions for maintenance, repair, and disposal. The repairs are not always performed at the VM's facilities, so these instructions are written for a broader range of technicians, complete with warnings and symbols to call attention to critical safety instructions. Allowable software configurations are a special concern, since the hardware is the same. For example, electric steering and stability control is specially tuned for each vehicle configuration and content. Normally, diagnostic codes are provided to aid in the repair and are referenced in the maintenance manual. Specific instructions are provided for safe maintenance. Systems that are not deactivated are specified: for example, stability control is not allowed to be disabled due to regulatory requirements. Safe replacement parts are specified: for example, certain aftermarket parts are not recommended for some systems or configurations. Often, repairs are monitored by tracking field service codes used for diagnosis by maintenance personnel as well as by other field-monitoring methods. Instructions for disposal are also broadly communicated. Both maintenance and disposal require special tools: for example, airbag deployment requires particular tools. Warnings are included for required precautions, for safety reasons.

It is common automotive practice to include extensive information in a user's manual. Often a quick guide is also included, and the extensive manual is indexed. Features, what to expect, warnings, and driver responsibilities are all included. Warnings include, for example, limitations of the driver-assistance feature due to weather conditions. It is not unusual for these warnings to be supplemented and reinforced by display messages or other driver warnings such as sounds, lights, and symbols. For example, a warning sound and image may be provided if the driver is not holding the steering wheel and the lane-assist system is engaged; this occurs several seconds before the system is disabled. In some cases of automated driving systems where the driver is expected to take over in case of a failure of the automated driving system, a driver monitor is used: for example, sensors in the steering wheel or even a camera system that monitors cues of driver alertness. If needed, an alert is provided, and the driver recovers.

Warnings in the user's manual provide information concerning necessary maintenance. Electronic systems supplement this by providing alerts if a sensor fails or, for example, if the steering system, stability control system, or airbag system senses a problem needing attention. If ignored, the systems are disabled. The user manual provides information about third-party accessories that interfere with vehicle features such as systems that use cameras, lidar, or radar. If sensor performance is degraded due to blockage, the system will not perform properly: an audible or visual alert results, to inform the driver.

Production Planning

In practice, the implementation of production planning is rigorous and audited. For example, when this author audited the PFMEA and specifications and walked the production line, he would check not only the existence of the specifications, but also whether they were maintained. The process was audited not only for the specified controls and capability, but also for the handling of rejected and reworked parts so they were not mixed. Test equipment was audited for maintenance, and practices for electrostatic discharge (ESD) were observed, including individual testing of the equipment prior to entering the production area. In general, the practices required by ISO 26262-7 appear to be consistent with common automotive practice. They were debated by automotive safety experts, and there seemed to be common agreement.

In addition, tier 1 suppliers often provide information to the VM concerning handling and installation of the items they provide. Special attention is paid to ensuring correct configurations and calibrations at the VM. Processes concerned with this are also jointly reviewed. Changes by both the tier 1 supplier and the VM must be carefully managed for compatibility. If there is a possibility that mishandling, such as dropping a component, could result in a safety issue, then the tier 1 supplier considers it a duty to warn the VM. This is also true for tier 2 suppliers. Electronic components are susceptible to ESD, but the failure occurs later, so warnings are provided. Similar attention is paid to disassembly and decommissioning. For example, special care must be taken when disassembling a steering column or steering wheel in the presence of an airbag system. The suppliers' behaviors result from impassioned concern about safety and pleasing their customers. Such behavior is consistent with ISO 26262-7; the industry is somewhat self-regulated.

ISO 26262-8 – Supporting Processes

ISO 26262-8 provides requirements for safety areas that are common for all phases of the safety lifecycle. While this avoids repetition, it also requires increased cross-referencing to understand all the requirements for each phase. Still, once the requirements of the supporting processes are understood, knowledge of all the applicable phases is enriched. Many of the requirements for supporting processes are achieved independently from the phase if the process is implemented at the project level. In practice, achieving compliance is a mixture of compliance at each phase and at the project level. Often, each domain, such as software and hardware development, has domain-level processes that support many projects. Variations on such an organization are not unusual in automotive practice; the project controls planning and milestones.

Distributed Developments

Distributed developments are common automotive practice. When a customer, such as a VM, evaluates potential suppliers and provides a request for quotation (RfQ), it often contains requirements to meet safety requirements of a specified ASIL, and may or may not identify those requirements. There are pre-sourcing audits, questionnaires, or both as an adjunct to the procurement process. The purchase order, on rare occasions, also contains a term or condition related to safety or ISO 26262. This implies a criterion for choosing a supplier that is based on the supplier's capability to meet requirements having a specific ASIL attribute. A proposed list of documents is exchanged, with responsibilities assigned for each. If there are issues with confidentiality, for example with the FMEA, these are negotiated, perhaps by agreeing to a joint review. Sometimes a supplier's internal audit reports are shared with the customer, although the customer will still schedule additional reviews that serve as an audit. In practice, it is not unusual for both the supplier and the customer to perform the HARA, or part of the HARA (except hazard analysis of application-specific features) before the DIA. The customer may specify that the customer will supply the HARA; if so, then the supplier usually resolves the differences with the customer: for example, by mapping the supplier's assumed safety goals to those of the customer. If the HARA differences cannot be resolved after review between the customer and supplier, the supplier includes a note in the safety case to explain the discoverable differences without prejudice. Such a note has value to both parties. If neither party has a preexisting HARA, then it is performed jointly or by a customer-specified party. At least some understanding or assumptions of the safety goals and ASILs are needed for the supplier to provide the quotation, including compliance with these safety requirements.

Once the DIA is agreed to, the parties then implement the agreed-on schedule. In practice, it is not unusual for the schedule to be heavily front-loaded in order to have the technical safety requirements and detailed safety requirements available by the design phase or earlier. This can be challenging unless most of the requirements are carried over from a previous application: increased resources are required at the project start instead of ramping up resource availability; and many requirements are derived from the safety analysis of the design, such as the design FMEA, SPFM, LFM, FTA, and software safety analysis. These requirements elicited in the design phase can be accommodated by updates of the requirements as the analyses are completed.

Some change requests and design modifications need to be managed during development. If the delay is too great due to availability of requirements, recovery planning is necessary during verification. To mitigate this risk, automotive suppliers often develop a SEooC even at the tier 1 level. It may have optional configurations to accommodate as many potential applications as possible, but with a head start on analyses, requirements, and verification applicable to the intended applications. This is especially true of tier 2 suppliers, such as silicon suppliers. The design and verification are time consuming, and predevelopment reduces schedule risk.

During development, communication is maintained between the customer safety manager and supplier safety manager. In practice, after the DIA is agreed to, such communication is less frequent. If there are issues on either side, they are communicated promptly and dealt with jointly. Both parties recognize the importance of this action. Also, the need for safety actions, and safety-related assumptions made concerning items outside the scope of the supplier's item, are communicated somewhat formally. This improves the safety case for both parties with evidence of communication and compliance. Safety-related assumptions become requirements. The compliance evidence results from verification.

While the functional safety assessment in a DIA can be performed by either party, according to ISO 26262-8, in practice the evidence is accumulated progressively during the project for reference by the supplier in a document that provides the argumentation for the safety case. If the supplier has a safety organization that includes an assessor and auditor for the project, it is reviewed and approved by them before being presented to the customer. In some cases, this is done progressively during development, but these reports are at most interim reports until all the evidence is available. Evidence of production, field monitoring, and disposal safety is part of the safety case as required. The customer participates actively using interim audits and schedules a final audit to review the safety case. Sometimes a third-party auditor is used. This audit report becomes evidence.

Requirements

ISO 26262-8 has requirements for safety requirements. In automotive practice, management of requirements has become a priority in product development. Part of this tendency in automotive practice toward a culture of requirements-based development is the priority given to automotive SPICE maturity levels as a condition of business awards. When teaching an introduction to ISO 26262 to students from six continents, this author emphasized that "It is all about requirements." It is not unusual for system requirements to be reviewed by customers separately from safety requirements and for the customer to use different people to review them. The tools for managing requirements are usually the same for safety-related requirements and other functional requirements; they are model based, but usually the requirements are in a database for ease of management. One of the issues, therefore, is that ISO 26262 has different required attributes for the requirements: a unique identifier, status, and ASIL. As the requirements database is shared across a global enterprise and selected and managed by non-safety personnel, ensuring the availability of these attributes is challenging. Nevertheless, the automotive industry is complying. Usually, every safety audit checks requirements traceability, which drives compliance.

Configuration Management

Configuration management of the item and of documentation is common practice in the automotive industry. It is necessary to be able to retrieve the documentation for a product build, including software, many years after it has been launched for production and public sale. Therefore, automotive practice is often to include safety case documents, such as work products and analyses, in the documentation system and workflow being used. Since some of the safety case documents require independent assessment, this can cause an issue with modifying the process just for these work products. Still, the automotive industry manages this internally and for customer review. Sometimes special tools are used to management the configuration of safety work products; these tools include facilities to maintain or reference a model and to link analyses for more efficient execution. However, the work products are archived centrally to conveniently support each application. For some tools, documents used for work products can be autogenerated from the model and analyses.

Every automotive product development has a method to manage changes. ISO 26262-8 requires that these changes also consider the impact on safety and the effect on safety work products, and that they are reviewed by the safety manager. In practice, this causes a change in the traditional change management flow, which did not require this review but considered the impact on safety – for example, the FMEA. Some projects have the safety manager review all changes or have someone else in the workflow notify the safety manager when appropriate. Ensuring that safety is not compromised by product changes is a high priority in the automotive industry. Workflows are modified for efficient compliance, and the safety case includes evidence.

Verification

ISO 26262-8 discusses verifying work products using different methods including testing, and planning verification in every phase. In practice, verification of work products depends on a strong process for review of documents as appropriate. It is common automotive practice to perform verification tests on the B-samples (design-intent samples), and to verify the software and hardware separately as well. The verification of the work products is performed by domain experts prior to submission for independent assessment. This requires discipline, as there is schedule pressure from the customer's DIA as well as internal project metrics. Rework occurs after verification and assessment, and this is a schedule issue. Allocating time for the assessment queue and rework causes a shortened verification opportunity. For verification of requirements, automotive practice relies heavily on testing using test cases traceable to the requirements. Automotive practice has improved over the years in this respect due to improved engineering quality support as well as internal and external safety audits. These audits must always check requirements verification: this is critical to safety.

Documentation

There are requirements for documentation in ISO 26262-8, which go hand and hand with the requirements for configuration management. Automotive practice is to maintain a documentation process that supports the retention policy for each enterprise. Together with the configuration management process – for example, a software configuration management process using a specific tool – clear documentation can be available for every product launched, including drawings, specifications, and safety work products such as safety analyses. This documentation supports the safety case. Documents can be referenced in a summary safety case document with safety arguments, and later retrieved to provide the complete evidence needed. The change history of each document can provide evidence that the safety case was updated when there were changes to the product, such as updates to deployed software. This can be useful when there are successive vehicle launches with minor changes to an item, such as a stability control system. The changes are approved in each affected document, which supports the safety case.

Tool Classification and Qualification

Automotive development commonly uses a significant number of software tools. These range in scope from simulation tools that also generate application software code to spreadsheets used for safety analyses. ISO 26262-8 has requirements for classifying these tools and for qualifying them if needed. To avoid the expense of tool qualification, many development processes are used that can detect an error that is inserted into the product by the tool and potentially cause a violation of a safety requirement. For example, in model-based development, the modeling tool is qualified because it was developed to a standard and, if it is used in accordance with the manual, the output code is trusted. Then, if test vectors are applied to the embedded code based on coverage obtained on the model, the compiler does not need to be qualified. This allows the use of the latest processors and ensures that any errors will be detected.

For simpler tools, qualification can be performed by the developer by running test cases that exercise the tool. Often, in-house qualification is performed by a central organization in the enterprise so that qualified tools are disseminated. This avoids the expense and scheduling loss of each project performing the required tool qualification. However, it may be difficult to ensure common usage. Still, a central repository for qualification of tools that can be used is a competitive automotive process, like having an independent central safety auditing and assessment group; it avoids redundant work. But some complex tools are qualified externally, or acquired with certification of qualification, to simplify use across the enterprise and provide tool commonality.

Qualified Components

Sometimes in automotive development it is more efficient to use qualified software components for a specific purpose, such as an operating system. In practice, when this is done, it is necessary for the runtime environment to be as specified for qualified use of the software. In addition, the compiler used for the qualified software needs to be the same as that used in the application on which it is to be used. This is all stated in the specification for the qualified software. In general, qualification is by a third party that complies with the requirements of ISO 26262-8. Suppliers of systems do not usually qualify software themselves: they usually purchase qualified software, which supports the safety case.

The first edition of ISO 26262-8 was somewhat ambiguous in presenting the requirements for qualification of hardware components. It was clear that class I components, such as capacitors and resistors, required no special safety qualification beyond common automotive quality practices. The safety aspects were handled by complying with ISO 26262-5 for the item in which they were used. Class II and class III devices were somewhat more difficult to resolve. In the second edition of ISO 26262-8, the clarity of the requirements for the qualification of hardware components has been substantially improved. There was broad participation of tier 3 electronic component suppliers as well as participation of tier 2 suppliers in the redrafting of these requirements. As a result, the qualification of class II and class III components is more broadly employed in automotive practice. The requirements governing classification are clearer, and qualification is more straightforward.

It is common automotive practice to qualify components such as sensors when they are going to be used in multiple developments for multiple applications. Some tier 1 suppliers continuously improve and update components to maintain state-of-the-art efficiency and performance. These components are then qualified in accordance with ISO 26262-8 and have enough evidence of safety to be used in their intended applications. This reduces overall resource consumption for the enterprise without compromising safety. In some case, a class III component such as a commercial microprocessor without a safety manual is used on a specific project. The second edition of ISO 26262-8 provides increased clarity about the evidence and justification required for this situation and notes that this is not the preferred approach. In automotive practice, SEooCs are widely available; this is a preferred approach.

Proven-in-Use

The proven-in-use argument is not as widely used in automotive practice as expected. The service period for proven-in-use credit in ISO 26262-8 is 10 times longer than required for IEC 61508. This service period was determined during discussions prior to the release of the first edition as necessary to avoid granting proven-in-use status to candidates that could have caused a safety recall. Because of this, it can be difficult for a VM to take advantage of proven-in-use credit. A tier 1 supplier can take proven-in-use credit for a basic legacy product that has provided service on multiple launches without change. For example, a basic electric or electrohydraulic product may not have a prepared safety case but may have been used without change or an uncorrected incident; the service hours are counted from all launches, and proven-in-use credit is taken.

The proven-in-use candidate does not need to be an entire vehicle or item: it can be an element of an item, such as unchanged software or hardware. However, the automotive industry is very cost competitive, so improvements in nearly everything are needed in order for VMs and suppliers to retain market share. As a result, most hardware is changed to improve performance at a lower cost. The improved performance and lower cost constrain the software, and the software must support the improved performance. Thus, change is prevalent throughout the automotive business, and service time without change is less than needed to comply with proven-in-use requirements. Change is managed and embraced – few elements remain unchanged.

T&B

Some suppliers and truck manufacturers chose to apply ISO 26262 : 2011 to items that were integrated into vehicles. Some manufacturers were already using ISO 26262 for other automotive products, and the internal infrastructure already existed. In some cases, automotive items developed in accordance with ISO 26262 were adapted to a truck application. During the development of the second edition, there was broad participation from the T&B industry. Requirements are included in ISO 26262-8 for interfacing an item developed according to ISO 26262 to a complete vehicle that is out of scope of ISO 26262. Also, requirements for interfacing systems not developed to ISO 26262 to a vehicle developed to ISO 26262 are provided, along with requirements for interfacing systems not compliant with ISO 26262. It can be expected that compliance with ISO 26262 will broaden in the T&B industry, now that ISO 26262 addresses T&B needs; such use has already started.

ISO 26262-9 – Automotive Safety Integrity Level (ASIL)-Oriented and Safety-Oriented Analyses

ASIL Decomposition and Coexistence

ASIL decomposition, the decomposition of requirements with respect to ASIL, is a common technique used in the automotive industry. Requirements and limitations of this technique are described in ISO 26262-9. While this technique does not apply to the hardware safety metrics of ISO 26262-5, there can be significant benefits when it is applied to software requirements. Safety goals are always assigned an ASIL, and this ASIL flows down to the functional safety concept. Here ASIL decomposition can be applied, for example, by assigning functional requirements to one item on the vehicle with a lower ASIL or QM (denoted, for example, QM(D)) and detection requirements (denoted, for example, D(D)) to another item on the vehicle with a higher ASIL. This can reduce the cost to the VM while not compromising safety. Also, as the requirements are further decomposed into technical safety requirements for hardware and software, the opportunity to perform further ASIL decomposition is available. The requirements assigned to software are often decomposed with respect to ASIL, which drives some independence in the software architecture. Often this independence is supported by hardware features in the target processor, such as memory partitioning. There is a significant opportunity for software operating systems that support the independence needed for ASIL decomposition. The operating system can prevent some dependent failures from diminishing the independence needed in the software architecture to support ASIL decomposition. For example, logical and temporal sequence monitoring is provided with minimum resource usage. While many decompositions can be used, perhaps the most advantageous is QM(X)+X(X), such as QM(D) + D(D), which allows functions to be developed to QM(X) and safety mechanisms to be developed to X(X). Test cases for safety mechanisms are simpler and verification is more straightforward.

Sometimes, when legacy code is used extensively for an application, the criteria for coexistence of ISO 26262-9 are not met. Some code provides functions that have a lower ASIL requirement than the primary function, such as a tire-pressure-monitoring function, based on a wheel speed algorithm, embedded in a stability control braking system. Because the tire-pressure-monitoring function is intermingled with code that fulfills higher ASIL requirements, like directional control using several signals including wheel speeds, all the code is developed to the higher ASIL requirements. In other cases, the criteria for coexistence are fulfilled. Consider a new stability control system that contains some unrelated customer code for a QM function. This customer code is partitioned in memory and executed in protected time slots that cannot interfere with the time slots dedicated to the code meeting the higher ASIL requirements, and does not provide any signals to be processed by the higher-ASIL code. Noninterference in complying with the higher ASIL requirements is completely ensured, and the requirements for coexistence are fulfilled. This is not unusual automotive practice.

Dependent Failure Analysis

In composing ISO 26262-9, the committee put special emphasis on improving the clarity of DFA. The importance of having an argument for avoiding dependent failures has increased significantly in the automotive industry. Part of the reason for this increased emphasis is the anticipation of fail-operational systems moving into the space of fail-silent systems. The systems need to continue to operate at some levels of automated driving systems, perhaps even at a lower level of automated driving, until a distracted or otherwise disengaged driver takes over. In some cases, this could be true for the entire journey. To enable this capability, some level of redundancy is anticipated, and this redundancy requires evidence of independence. Dependent-failure analysis provides evidence.

Further driving the emphasis on DFA is the increase of integration. Many of the examples in ISO 26262-9 use such integrated circuit elements. It is assumed that an audit, either internal or external, of an automotive product that includes integration will have its DFA closely scrutinized for completeness and correctness. To manage this, many suppliers, as well as consultants, have developed systematic methods to execute a DFA. Many times, the design FMEA anticipates cascading failures in hardware. Common-cause failures are addressed by matching the members of FTA cut sets with the causes in the FMEA to see if unity cut sets result. There are conference workshops to train how to perform a DFA, some of which emphasize software. The use of checklists is proposed by ISO 26262-9, and an annex is provided for guidance.

Other Analysis

Other analyses beyond DFA are common automotive practice. ISO 26262-9 discusses qualitative analyses such as FMEAs, HAZOPs, FTAs, and ETAs, all of which are commonly used in automotive practice. Quantitative examples are also given, but the emphasis here is on reliability in actual practice. The architectural metrics of ISO 26262-5 are always performed. Other quantitative analysis is not as widespread, but reliability is always ensured.

Partially because of the existence of safety goals in ISO 26262, the emphasis on deductive analyses, such as an FTA, is widespread in automotive practice. Sometime these are quantitative, but that is not required by ISO 26262. The accuracy of the single-point and residual-failure-rate summation, as derived from the SPFM and LFM, is acceptably precise in automotive practice. A qualitative FTA is performed on software, or on a special aspect of software, such as communicating an incorrect value. Other software architectural analyses include timing and HAZOP analyses. The timing analysis is broad and precise, considering the worst case in order to ensure required performance. Notwithstanding this timing analysis, a HAZOP of the software architecture that examines each interfacing information flow includes guidewords such as *too early*, *too late*, *missing*, *incorrect*, and *too often*. There is guidance in annexes of ISO 26262-6 that are referenced in ISO 26262-9 for software safety analyses. Systematic software safety analysis provides greater confidence in the completeness of the software safety requirements, but it is a challenging in automotive practice because the software scope is extensive.

ISO 26262-10 – Guidelines on ISO 26262

ISO 26262-10 reviews key concepts that are broadly accepted across the automotive industry. These include the difference between the general standard, IEC 61508, and the automotive standard, ISO 26262. This guidance notes that the safety goals are mandatory, and that the hazard analysis does not imply that a failure will always lead to an accident in automotive practice. Unlike IEC 61508, ISO 26262 targets series production products. It includes managing safety across multiple organizations. Automotive practice has been to accept these differences even by those that have previously used IEC 61508; they are seen as improvements for automotive. IEC 61508 users participated in writing ISO 26262.

Use of Terms

The concepts of *item*, *system*, *element*, *component*, *hardware part*, and *software unit* are understood in automotive practice, but perhaps not generally used except when such vocabulary contributes to the clarity of communication. *System* is used with more latitude in practice. Likewise, the taxonomy of *fault-error-failure* is generally understood in automotive practice and commonly used. However, the precision is not rigorous in all cases; precise usage is used in analyses.

ISO 26262-10 explains that the FTTI only exists at the item level and can be included as part of a safety goal. Including the FTTI as a part of the safety goad, especially after hazard testing, is not uncommon in automotive practice. Nevertheless, the FTTI has been used at the element and component levels in the past. Fault-handling time is provided in the second edition of ISO 26262-1, as discussed; it enables more precision in specifications than the previous use of the FTTI at multiple levels. Likewise, emergency operating time is now available for use to describe the interval before a permanent safe state is reached, which allows further precision. ISO 26262-10 demonstrates this in a timing example using a valve. The safety-related time intervals are used, supporting precise usage.

ISO 26262-10 describes some aspects of safety management, including the interpretation of work products as a level of abstraction that shows compliance with the requirements of ISO 26262. This agrees with automotive practice, and the use of tools enables this view of work products. The tools provide the required evidence that is assessed with or without printed documents. The use of such tools is widespread in the automotive industry; in some cases, VMs and suppliers access the same tool with a protocol that mutually protects intellectual property (IP).

ISO 26262-10 also provides an extensive explanation of the differences between confirmation and verification. This is appropriate, as misunderstanding of these terms is not rare in the automotive industry. Safety practitioners insist that verification measures take place prior to confirmation reviews of work products in order to ensure that the evidence reviewed is technically correct before it is independently assessed for conformance to ISO 26262. Domain experts prefer the assessment before submitting the document to be reviewed technically because of the risk that changes needed to pass the assessment will require re-review. A further misunderstanding is that if a work product requires assessment, then a verification review is redundant. ISO 26262-10 targets resolving such misconceptions.

Safety Assessment

The discussion of the functional safety assessment in ISO 26262-10 addresses many of the practices used in the automotive industry. The explanation demonstrates that the practice of having an enterprise process that is periodically audited can support the efficiency of the functional safety assessment prior to launch. Also, ISO 26262-10 states that the safety expert who serves the function of the independent assessor of the work product is also the person who conducts the functional safety assessment. This is an automotive practice when there is a central assessment organization as well as when an independent consultant is supporting a project. The limited scope of assessment throughout the automotive supply chain also agrees with automotive practice. It is not unusual for the safety case of a tier 1 automotive supplier to reference safety assessments or the safety manual of tier 2 suppliers. Ultimately, the functional safety assessments roll up to the VM, which may also perform assessment. The compilation of evidence ends there.

The elements of the safety case are discussed in ISO 26262-10. These are explained as the safety goals, evidence of compliance, and the safety argument. Early in the development of the first edition of ISO 26262, this was debated before it was finally agreed that just the compilation of work products was not enough: the argument for why the accumulated evidence demonstrated safety was also needed. Automotive practice is to have a statement that the safety case adequately supports no unreasonable risk for series production and public sale; this statement can be referenced for product release. ISO 26262 requires independent acceptance of this argument.

Classifying and Combining Safety Goals

ISO 26262-10 provides examples of determining and classifying hazards and combining safety goals. In practice, these examples are simple and straightforward compared to the extensive effort that is expended in determining scenarios and the potential hazards in each of them. The use of traffic statistics is mentioned, but the effort needed to analyze available accident data to determine severity is not shown. Combining of safety goals is shown: in automotive practice, the supplier and VM may complete an extensive HAZOP-based analysis that results in a multitude of different combinations of safety goals with diverse ASILs assigned. All these different safety goals and ASILs need to be resolved to support collaboration as well as accurate quotations; this is a typical, expected concept-phase activity.

Requirements

The flow of safety requirements to hardware and software is illustrated in ISO 26262-10, along with verification. Automotive practice generally follows this flow with respect to safety requirements. However, there is substantially more interaction than is illustrated because of schedule pressure that drives suppliers to develop the technology as well as to start development of the intended item out of the context of a vehicle contract. After award of the contract, the flow is interrupted as new input from the actual application becomes available, including requirements for prototype samples in parallel with the production-intent design. Hardware in samples that are used more than a year before production is not identical to the hardware in the production design; likewise, software in prototypes may differ from the production-intent software years later. Nevertheless, the flow of requirements in practice generally agrees with the flow of requirements illustrated in ISO 26262-10 for the production-intent design. At each step in the flow, they may undergo refinement through collaboration with the application customer. Further additions come from the assumptions of others, such as an SEooC microprocessor, as well as requirements derived from analyses. All these requirements need verification.

Failure Rates

ISO 26262-10 provides considerable discussion and explanations concerning failure rates of components and their classification for calculations and summations that are used in determining the architectural metrics in ISO 26262-5. This extensive discussion is appropriate for automotive practice because considerable effort is expended determining these hardware metrics. Considerable effort is also expended by suppliers presenting these metrics to customers, and by customers and safety assessors reviewing these metrics.

In addition, ISO 26262-10 incudes discussion of determining the average failure rate during the life of the application. While this effort may seem excessive due to the uncertain precision of estimating failure rates, and the normally high quality of automotive electronic components, it is warranted because the metric analyses elicit detailed requirements for safety mechanisms. Eliciting and verifying these requirements is essential to ensuring the safety of the item as well as avoiding the expense of recalls. These requirements for safety mechanisms may also include requirements for the software reaction to the exception of a hardware failure. Safety mechanisms for potential memory failures and arithmetic logic units are considered, and the requirements may flow to software. Metrics analyses support completeness.

SEooC

The discussion of a SEooC in ISO 26262-10 is a true description of automotive practice. Significant development time and resources can be saved on an application project through use of SEooC. In the case of an SEooC as a system, development of the SEooC includes a concept phase where a hazard and risk assessment is performed based on the assumed effects at the vehicle level. Safety goals are determined with ASILs and flowed to the system level to determine assumed functional safety requirements: including, for example, assumed requirements from other items and assumed requirements on other items. Then these are further detailed into technical safety requirements, including requirements derived from safety analyses and top-level software and hardware requirements. These requirements also include any other assumed requirements on other systems and technologies that are then listed in the safety manual. The technical safety requirements flow further to detailed hardware and software requirements. All these requirements are verified.

Many automotive SEooCs are hardware, such as microprocessors. These assume no safety goals but do assume hardware requirements with ASILs. Safety analyses elicit further requirements, many of which become assumptions on the system software. The hardware requirements are verified, and the safety manual lists the assumptions. It is common automotive practice to meet these requirements: the project using the SEooC checks the assumptions, and if any are not met, analysis determines whether they are needed. If so, then changes must be made by the project using the SEooC or to the SEooC, as explained in ISO 26262-10.

Proven-in-Use

ISO 26262-10 provides a plausible example of a proven-in-use argument for hardware to be reused. This argument is consistent with the practice of establishing proven-in-use credit in the automotive industry. However, if a software change is only a calibration change, an impact analysis can be performed to see if such a change is safety-related, particularly if the software was designed to be safely recalibrated and has been safely recalibrated in the past. The practice of obtaining proven-in-use credit when carrying over hardware or software is somewhat rare in the automotive industry because of the constant changes made for improved performance and cost reduction. It can occur when reusing legacy products.

ASIL Decomposition

The example of ASIL decomposition provided by ISO 26262-10 primarily discusses the allocation of requirements to hardware elements. This is a realistic example especially because it requires the addition of hardware. Automotive practice is to add such hardware when needed, particularly in the context of switches or switch components that may fail. The application of ASIL decomposition is more commonly applied in software, even without additional hardware. The motivation here is not reliability but efficiency of implementation and verification. More complex functionality implemented in software is a lower ASIL or QM, and simpler monitoring software is a higher ASIL. Requirements for independence are needed, and compliance is often supported by hardware partitions as well as the operating system to ensure spatial, temporal, and logical independence. Then compliance can be achieved more simply and verification is more straightforward.

Fault Tolerance

The discussion was extensive before arriving at a consensus concerning the guidance provided by ISO 26262-10 for fault-tolerant items. The guidance at the time of this discussion had to be forward-looking for various levels of automated driving features. Correct reference to the time intervals as well as how the intervals were to be used in the examples were topics for debate and agreement. In the end, the explanations agree reasonably well with automotive practice. When a system is to be fault-tolerant, agreement on what faults are to be tolerated is always needed. For example, in a steering system, faults of a torque sensor, electronic control unit, or motor need to be tolerated. Failures of the rack and pinion do not require a fault-tolerant strategy; they are simply avoided. As discussed in ISO 26262-10, agreement on emergency operation is needed, as well as the transition and warning strategy. For example, if the steering provides a reduced assist when the vehicle is moving, and at a very low parking speed, there is not a hazard. This is exactly like the generalized example in ISO 26262-10. On the other hand, another system provides redundancy when moving, and a transition is defined in the strategy. The ISO 26262-10 discussions were spirited with respect to the timing diagram: the result is a precise use of terms in an example that supports automated driving safety, as well as improvements in the fault tolerance of non-automated items. There is also discussion of ASIL decomposition and software.

The list of considerations includes the safe state to be reached. In some cases, this is after repair. When considering a level 4 automated system, a minimum-risk condition needs to be defined as a safe state, such as reducing the speed to the point that no hazard exists. This is shown in an example in ISO 26262-10. The need to do this may result from reaching the end of the EOTTI. ISO 26262-10 discusses how to determine the EOTTI using methods that can be used in automotive practice. When a portion of the item is disabled, the PMHF may change, and the lifetime is scaled to determine the EOTTI, for example. Some margin is provided, based on judgment: several journeys are chosen, and then the system is locked.

The use of ASIL decomposition is not as common, because the ASIL is dynamic after the transition to emergency operation. Nevertheless, it is possible, because the exposure and perhaps severity of the hazard may change. However, when the requirement of fault tolerance involves another item, these assumptions become requirements that may benefit from a decomposed ASIL attribute if the items are sufficiently independent. This is expected in the vehicle architecture selected to accommodate the automated features being included. The VM can ensure safe operation efficiently, and safety features become less costly.

Fault-tolerant software is mentioned in ISO 26262-10, and ISO 26262-6 is referenced. In practice, defensive programming techniques are used, and software quality assurance is employed and audited. This helps reduce the occurrence of systematic errors in software. Further assurance may result from using diverse software or from rigorous DFA and cross monitoring of similar software. Either practice, or both, can be resource intensive to implement. Nevertheless, the demands of automated driving demand this rigor: it is a focus of safety audits.

Tool Classification

A flow chart showing how tool error detection can be used to avoid the need to qualify a software tool is provided in ISO 26262-10. This practice is followed in automotive practice in some cases; for example, if test vectors derived from a model used to generate application software are used to verify the embedded code, then errors in the compiler are detected. This may justify tool classification 1 (TCL1) for the compiler, avoiding the need for its qualification. In automotive practice, when tools that are used in the enterprise by many applications that do not have enough error-detection methods embedded in their processes to justify TCL1, qualification is chosen. Then, when used consistently with the tool's qualification conditions, requalification for each project is not necessary.

Critical Characteristics

Special and critical characteristics are discussed in ISO 26262-10. The steps for determining special characteristics, implementing controls, and verifying that special and critical characteristics are achieved agree with automotive practice. Often, the dFMEA is the analysis of choice to determine design-related special and critical characteristic candidates. Then the characteristics confirmed as special or critical are included on the drawing or other design document provided to production process and quality personnel, who determine how to implement the capability to achieve these characteristics. Often, the enterprise has policies deployed to ensure consistent rigor in this regard: a mistake-proofing process, a statistical control, or a screening such as a 100% test. These steps are common automotive practice and are broadly audited.

ISO 26262-10 provides some guidance on an FTA as well as the FMEA. It states that an FMEA deals only with one failures at a time. This is normally the case in automotive practice, though in some fail-operational systems, the effect of a second failure while in the safe state is also addressed in an FMEA and leads to increased complexity in the analysis; it can be accomplished in one FMEA or two (or more) FMEAs. Similarly, in an FTA, this can be accomplished using house events and AND gates to switch between the following modes of operation: startup, normal operation, emergency operation, degraded operation, safe state, and shutdown. Some suppliers also use house events to include or exclude certain features of the item in the FTA, or to switch among customers for consistency. While different levels of granularity are included, it is not unusual in automotive practice for the FTA to be somewhat hierarchical and go to the level of the FMEA for base events. FTA theory guides the FTA analyst to consider immediate causes, and the more detailed the causes that are specified, the more granular the FTA becomes. For example, a motor failing to turn is caused by increased rotational resistance to the rotor, or insufficient electromotive force applied to the rotor. Alternatively, this can be specified as a blocked rotor or shorted windings. While the first set of causes leads to the same causes as the alternative, the first also leads to others, such as increased air gap or demagnetization of the magnets. In automotive practice, both are found. The alternative approach is somewhat like a reverse FMEA and is sufficient for many FTAs. The faults and failures are considered in combination. While ISO 26262-10 discusses combining an FTA with the FMEA for the system, note that not all failures from the FMEA are included in the FTA – only those that lead to the top event. Still, the combination helps with completeness, and the approaches are complementary.

ISO 26262-11 – Guidelines on Application of ISO 26262 to Semiconductors

Background

In the first edition of ISO 26262, there was not specific guidance for semiconductor components such as application specific integrated circuits (ASICs), microprocessors, or sensors. Very soon after publication, a multi-part publicly available specification (PAS) was created and used as practical guidance for the application of ISO 26262 to semiconductors. This PAS was then the basis for clause 13 of ISO 26262-8 and the new part, ISO 26262-11.

There was significant global collaboration to develop the PAS and to implement the subsequent inclusion in the second edition. ISO 26262-11 is expected to be followed in automotive practice: it was developed out of an automotive need for guidance, and it replaces the PAS.

The stated intent is to provide guidance for development of a semiconductor SEooC that can also be adapted to semiconductor components in the context of an item. In the case of an SEooC, development is based on assumptions about hardware requirements that potentially support an item, as well as assumption about the potential item to support the SEooC. In the case of a semiconductor developed in the context of an item, the hardware requirements are inherited from the item. In actual automotive practice, an ASIC is developed out of the context of a specific item while the ASIC developer is cooperating with the item developer that plans to use the ASIC on several applications for vehicles. This allows the ASIC supplier the necessary time for development, which may exceed the time available in item development. Likewise, an SEooC semiconductor may be in development with a semiconductor developer that also cooperates with one or more item developers to establish the assumed hardware requirements. Generally, the intent agrees with automotive practice, although practical consideration causes exceptions.

In automotive practice, various levels of granularity exist in the analysis of semiconductors. There have been cases when very granular analysis is performed at the elementary subpart level on some parts of the semiconductor to examine possible failure modes to ensure detection or to assess potential common cause or other dependent failures. In other cases, as noted in ISO 26262-11, such detailed analysis is not needed because the higher-level failure mitigation is assured, and the probability of common-cause failure is sufficiently remote by design. Still, in practice, there are different biases between analysts. ISO 26262-11 is intended to commonize the approach by providing uniform guidance.

The application of semiconductor internal failure modes to the system level is described in ISO 26262-11 in a manner generally used in automotive practice. The semiconductor is handled separately – for example, as an SEooC – and the higher-level failure modes are included in the system safety analyses. While this seems intuitive, verifying the semiconductor analyses with the tier 1 supplier can be difficult. The semiconductor analysis is extensive and hierarchical, as described in ISO 26262-11. The tier 1 supplier may wish to examine it deeply enough to have confidence and perhaps link it to the system analysis directly. This is impractical for the semiconductor supplier. The guidance provided in ISO 26262-11 should commonize the level of analysis by the semiconductor supplier and the acceptance criteria of the tier 1 suppliers. This is expected to support shared analysis, because the standard requires agreement.

IP

ISO 26262-11 includes a detailed discussion of IP. In this discussion, IP is limited to hardware representations that are included in semiconductors, such as cores or buses. The representations are the physical representation or a model that is implemented by an IP integrator. The discussion provides detailed guidance for the interplay between IP suppliers and integrators and is primarily from the semiconductor or IP supplier's point of view, as opposed to a VM or tier 1 supplier's perspective. Nevertheless, the provided guidance calibrates the expectations of VMs and tier 1 suppliers, which tends to drive expectations in the automotive industry since VMs, tier 1 suppliers, and component suppliers reached a consensus on the guidance. The guidance is not normative, but compliance is still expected.

The guidance applies the standard to IP suppliers and integrators in a manner similar to the relationship of customer and supplier elsewhere in the standard. IP is described as being provided as an SEooC, in context development, in a hardware component that is qualified, or in a proven-in-use candidate. Assumptions are treated as requirements on the other party. If compliance with these assumptions is not possible, changes are to be managed by change requests to either party. In automotive practice, such process control is expected and audited by a third party if the IP integrator is providing an SEooC, or by the customer's component engineers or safety managers if the IP integrator is developing the component in context. Particular attention is focused on hardware metrics and safety mechanisms. Obligations of the IP supplier and integrator are described in ISO 26262-11 to ensure that diagnostic coverage is provided and verified after integration. Initial assumptions are verified collaboratively, and potential issues are resolved.

Systematic Faults

Control of potential systematic faults is also addressed by ISO 26262-11, in the context of the IP lifecycle. This lifecycle is compatible with that followed by the automotive industry, so it is expected to be questioned and audited for evidence of compliance by automotive customers of safety-related semiconductors. A safety plan is required, and it is tailored to include the hardware lifecycle because this is appropriate for semiconductors whether developed in or out of context. The only distinction is that in context, the requirements are known; while out of context, the assumptions are treated as requirements. This is true for the IP supplier to the IP integrator just as for the semiconductor supplier to its customers. It is to be expected in the automotive industry for integrated circuits used in safety-related products.

Further, any calibrations or application-specific features that are configurable require evidence of compliance with safety-related requirements, and analysis of dependent failures is performed. This guidance is expected to be supported by automotive customers of semiconductor manufacturers: they have safety managers or component engineers who are conversant in the guidance provided by ISO 26262-11 that is relevant to their application, and evidence that the guidance is followed is part of their safety case. Review of the verification report and safety analysis report is expected by the IP integrator, and evidence of adequate review is expected by customers of the Integrated circuit. Evidence of confirmation measures, such as the report of an independent safety assessment of the semiconductor development process including the integration of IP and any work products exchanged, is also expected. A list is provided in ISO 26262-11 of the potential documents in a potential documentation set, which can be tailored for IP. After integration of the IP, testing is also documented.

Integration of black-box IP is also included in the guidance in ISO 26262-11. Such black-box IP is widespread in the automotive industry due to integration of customer and competitor IP. Assumptions need to be provided by the IP supplier, along with verification methods. Because transparency is lacking, the IP supplier also needs to provide certification or some other kind of evidence of compliance with ISO 26262. Otherwise, another argument for compliance is referenced, such as compliance with another standard or some other appropriate justification. This is referenced in the safety case, along with verification.

Failure Rate

There was high interest and participation in providing base-failure-rate guidance for semiconductor devices in ISO 26262-11. This was partially the result of an overly pessimistic failure rate being used, perhaps based on standards that did not represent the actual failure rates of semiconductors observed in the field or by reliability testing by semiconductor manufacturers. General information on base-failure-rate estimation is included in ISO 26262-11, along with extensive coverage given to the calculation of base failure rates based on a standard equation. In automotive practice, this emphasis is justified because nearly every automotive safety-related application that includes electronics requires a base failure rate for semiconductor components. During discussions while drafting the second edition of ISO 26262-5, it was agreed to allow the use of field and supplier data with a 70% confidence when computing architectural metrics and PMHF. The information, tables, and guidance in ISO 26262-11 provide enough information to accomplish this even in complex mixed-technology semiconductor devices. It can be expected that these methods will be employed broadly in the automotive industry. Component engineers and safety engineers at tier 1 suppliers as well as VMs review the use of these techniques and exchange documentation, which supports the safety case.

Transient faults are also discussed in ISO 26262-11. Interest in transient faults continues to increase in the automotive industry as semiconductor technology advances. Transient faults are distinguished from permanent faults caused by transients, such as damage from electromagnetic interference. The semiconductor supplier may have estimates or test data that support the transient fault rate that the supplier provides for a specific integrated circuit, and customers can review these estimates. Architectural techniques are used to provide logical and spatial separation to reduce the vulnerability to multiple-bit upsets, and these measures are expected to continue.

The failure rate of semiconductor packages is also discussed in ISO 26262-11, including the failure rate of all internal connections of the package, such as connections to the lead frame from the silicon die. Also, failures due to die attachment and failures due to encapsulation are included in the package failure rate. Failures to the printed circuit board that are external to the package are considered separately by the customer because they are process dependent, though ISO 26262-11 has notes that indicate some disagreement among references. In automotive practice, it is expected that package failure rates are taken into account as suggested by ISO 126262-11. They are considered separately from the die, and soldering is the customer's responsibility.

ISO 26262-11 is thorough in its examples and alternative methods of estimating the base failure rate of a semiconductor product. It pulls together the different types of circuit families and other references in a useful and organized manner. The effect of the mission profile, including power-up and power-down cycles, is discussed. Different standards are referenced for further consideration and comparison, and each term in the standard equation is developed. There is a discussion of failure-rate distribution based on area as well as other methods; and caution is given regarding multichip modules. Calculations based on mission profile and number of devices in a semiconductor device were included in automotive practice long before ISO 26262 was published. ISO 26262-11 represents the consensus of experts currently and is considered a reference when defining the state of the art, and therefore, its guidance is followed globally.

Dependent Failure Analysis

ISO 26262-11 provides guidance for DFA. Dependent failures are probably one of the greatest concerns related to the use of increasingly comprehensive integration of safety-related functions on semiconductor circuits in the automotive industry. This is especially true when on-chip redundancy is used in critical applications. It is expected that this concern will continue as applications include increased automation levels and greater frequency of fail-operational applications. In the guidance, the basic notion of dependent failures is provided, including common cause and common mode. Dependent failures have long been addressed in automotive practice. The concept of DFI is introduced, as well as coupling mechanisms. Previously in automotive, some checklists were used in DFA but matched the causes in an FMEA to the base events in an FTA to elicit potential unity cut sets after common-cause reduction was also used for common-cause analysis for discrete circuits. The systematic approach presented in ISO 26262-11 using a list of DFIs may become well accepted in automotive practice as it is also the suggestion of ISO 26262-9. The advice to proceed hierarchically can be followed as well as deciding the level of granularity that is acceptable. A workflow for DFA is provided. However, there are still concerns about coupling mechanisms present beneath the granularity chosen, leading to customers requesting deep dives into some more critical integrated elements. This possibility was not discussed in the DFI discussion of ISO 26262-11, but checking the sufficiency of information is in the workflow. Every step in the workflow is discussed in ISO 26262-11, and methods to mitigate the causes and effects of dependent failures are widely accepted in automotive practice. Concern about incremental growth of the integrated circuit failure rate due to additional on-chip hardware measures is an automotive consideration, as is the cost. Verification is also advised in ISO 26262-11, and several means are discussed. Having this apparent agreement encourages commonality, and acceptance is expected.

Production and Operation

ISO 26262-11 discusses compliance by semiconductor manufacturers with the requirements of ISO 26262-7. The conclusions are that compliance with these manufacturing requirements is achieved through compliance with quality standards and reusing the work products for evidence. ISO 26262-11 guidance is that the maintenance and disposal requirements of ISO 26262-7 are not applicable to semiconductor manufacturers. Automotive practice is to agree with this guidance. Semiconductor component-level repair is uncommon, so disposal restrictions do not apply.

Distributed Development

There is a relatively short section in ISO 26262-11 on interfaces in a distributed development with a semiconductor manufacturer. This is very common in automotive practice for tier 1 suppliers and semiconductor manufacturers, and the requirements of ISO 26262-8 are generally accepted as being applicable. ISO 26262-11 also provides guidance that there can be joint developments in which the semiconductor manufacturer is the customer and another party is the supplier. In this case, the requirements of ISO 26262-8 for the supplier are the obligation of the semiconductor manufacturer. When the supply is not safety-related, there are not obligations for a safety-related joint development. Automotive practice is to accept this; the guidance is not controversial.

Confirmation Measures

Confirmation measures are discussed in ISO 26262-11. This guidance is that confirmation measures and safety assessment are relevant to semiconductors whether considering a SEooC or IP. Confirmation measures considering safety at the item level are tailored away. Audits of the safety process and execution of the safety plan are performed, perhaps with a checklist. Confirmation reviews of the involved work products are performed, and any deficiencies are to be corrected. Such guidance is generally accepted in automotive practice. Tier 1 customers determine item safety relevance, and confirmation measures are reviewed.

Integration and Verification

ISO 26262-11 provides guidance for hardware integration and verification. Examples of how the relevant verification requirements of ISO 26262-5 are applied to semiconductors directly, even out of context, are provided. As these examples were reviewed not only with semiconductor providers, but also with VMs and tier 1 suppliers, it is expected that they will be accepted generally in automotive practice. For example, use of equivalence classes is applied to features. The guidance serves as somewhat of a tutorial on semiconductor practice for automotive safety practitioners. The guidance may serve as a bridge, and general automotive application is expected.

Digital components and memory, such as are found in microcontrollers, system on a chip (SoC) devices, and ASICs, are discussed in ISO 26262-11. There is extensive discussion of both permanent faults and transient faults. Both of these categories of digital faults are of high interest in the automotive industry with respect to safety-related products, so this guidance is relevant to help harmonize considerations. The disagreements that need to be harmonized relate to the level of abstraction that is appropriate to have the desired confidence in the analysis. Confidence centers on assurance that all the underlying technology faults can be mapped to one of the faults considered in an abstracted fault model. For example, the basis of the fault model for the digital elements is that the underlying technology faults will map to errors of omission, commission, timing, and value of the function. The analysis therefore will not extend into the underlying technology directly. Customers thus need to share the analyst's confidence in these assumptions in order to rely on the analysis to support the safety of the product that is intended to use the integrated circuit. ISO 26262-11 promotes this confidence in the guidance for the fault models by providing a second level of abstraction in the tables. This increases the transparency of the elements presented, but still does not delve too far into the underlying technology. It is expected that the guidance will be accepted by the automotive industry as a reasonable harmonization of the fault models considered and that it will lead to a practical level of analysis. Nevertheless, some component safety engineers will insist on a deeper dive into a sample of elements to ensure that the errors of omission and commission of a particular technology do not have a broader functional impact than that represented in the fault model guidance. Some suppliers improve transparency in other ways – customers always appreciate this, and it supports harmonization.

Analyses

Qualitative and quantitative analyses of digital semiconductor circuits are discussed in ISO 26262-11. Qualitative analysis guidance suggests this as the method that addresses DFA. This appears to agree with expected automotive practice. Only in rare cases is it expected to require quantification of a dependent failure of a digital integrated circuit. This occurs if the dependent failure appears to be unlikely, but evidence is needed to support that the probability of occurrence is low enough to be acceptable for an application; special measures ensure that such an occurrence is random and not due to a special cause, such as a manufacturing defect. Otherwise, qualitative analysis is enough and demonstrates dependent-failure detection.

Quantitative analysis is nearly always required for digital circuits, and the guidance provided agrees with expected automotive practice. Information is needed for at least the block-level structuring, and perhaps more to support the qualitative analysis of dependent failures. It can be expected that considerations of dependent failures and independence are taken into account early in the concept phase, as the guidance suggests. The level of detail is driven by the need to determine diagnostic coverage. For example, if the entire integrated circuit is duplicated on the same chip, dependent failures are adequately mitigated, and a comparison is performed externally, then a very limited level of detail is needed concerning failure modes in order to be acceptable in automotive practice. If the failure modes are addressed by more granular on-chip safety mechanisms or by dedicated software execution, a significantly granular analysis of failure modes would be necessary to justify quantitative diagnostic coverage that would be generally acceptable in expected automotive practice. The guidance should lead to this level of acceptability; a realistic estimate is usually preferred, and the analysis is conservative.

Faults

Guidance concerning treatment of transient faults is included in ISO 26262-11. Consideration of transient faults is a very relevant concern in the automotive industry. A qualitative discussion is enough, as covered in the guidance of ISO 26262-11. It is common automotive practice, in applications, to age faults and calculations either to average out potential noise or for to filter out potential false positives for detection. For example, before an intervention of a stability control system, wheel speeds, inertial sensors, vehicle speed, and steering angle are consistency checked several times. Such techniques mitigate the effect of many, if not all, types of transient faults. Thus, those areas that are still susceptible to failures due to transient faults can be the focus of quantitative analysis. For integrated circuits that are developed as a SEooC, this is not assumed unless it is listed as an assumption, which could put some limits on the applications that are not always acceptable. Architectural considerations are considered, as well as spatial and logical separation; sufficient information is needed for metrics computation and is expected by automotive practice.

ISO 26262-11 provides guidance concerning techniques or measures to avoid or detect systematic faults in digital design. This guidance concerns providing evidence elicited from a standard design process to demonstrate that sufficient measures were taken to prevent safety-related systematic errors in the final digital design. Guidance is provided for what should be included in this process. This approach will probably be generally acceptable since it is based on the requirements of ISO 26262-5. Some auditors look for evidence of sufficient analyses to elicit a complete set of requirements and evidence of compliance with these requirements by the design. This is indirectly addressed by verification of each semiconductor design activity; however, it is not explicitly stated. Similarly, it is suggested that software design techniques can be tailored to help prevent systematic faults at the register-transfer level of design. This argument is accepted in practice since all software design guidelines target control of systematic faults. Safety auditors nearly always audit requirements traceability. Nevertheless, the guidance has been generally agreed with in automotive.

Verification

ISO 26262-11 provides guidance concerning verification using fault-injection simulations. The guidance discusses using this method at the gate or register-transfer level to functionally verify the effect of stuck-at faults as well as verifying the effectiveness of associated safety mechanisms. Automotive practice has been to accept this verification method, and this may continue. It is a mature technique and is at a practical level of abstraction. For non-stuck-at faults, other techniques are suggested, and references are provided. Limitations are noted, but there is substantial literature that supports effective execution. Therefore, automotive industry acceptance is expected for these techniques for non-stuck-at faults.

The safety documentation set guidance for digital components is provided in ISO 26262-11. It includes the applicable safety plan and evidence to support that the safety plan was executed effectively. This evidence includes the specification, safety analyses, and verification. Any documents required by a DIA are also recommended, including the assumptions of use. Such documentation is provided in the automotive industry by suppliers of digital semiconductor suppliers, so the guidance in ISO 26262-11 makes this expectation a more transparent automotive practice that supports the customer's safety case. Guidance specific to digital integrated circuits is provided concerning safety mechanisms. Methods are listed with coverage rated low, medium, and high, and further detailed comments and descriptions are provided. In automotive practice, these will probably be accepted as 60% for low, 90% for medium, and 99% for high, unless further justification is provided. Such justification would probably be accepted, especially if it is mathematically derived or based on multiple safety mechanisms that do not fully overlap. Similar guidance has been accepted previously.

Analog Components

Analog and mixed-signal semiconductor safety guidance is provided by ISO 26262-11. The distinction made between the two is simple: if at least one analog element and one digital element are included, then the technology is mixed. This distinction is generally accepted in automotive practice. While the failure modes discussed depend on function, the distribution is stated to depend on implementation. The level of granularity needed will depend on the purpose of the analysis and concerns about dependent failures.

ISO 26262-11 provides a list of failure modes, and it is expected that they will be generally accepted in the automotive industry because they are at a practical and useful level. The discussion is on the level of detail needed to analyze the causes and acceptability of the design. Guidance provides details of the trade-offs for various levels of granularity for purposes of determining needed safety mechanisms as well as determining failure-mode distributions. Examples are provided; still, some discussion is expected to agree on the level of granularity that is needed to meet hardware metric requirements and provide confidence in the sufficiency of the evidence for the safety case. Toward this end, the guidance suggests that semiconductor suppliers provide the recommended distribution of failure modes based on the process technology employed. This is expected and accepted in the automotive industry, and the determination of safe faults is straightforward and consistent with current automotive practice. The examples provided are straightforward, and the reasoning is clear.

Guidance included in ISO 26262-11 for DFA of analog circuits expands on the general guidance provided for DFA of hardware elements of a semiconductor integrated circuit. Qualitative analysis is recommended, as is automotive practice for DFA. The guidance suggests that since analog circuits are more susceptible to noise and electromagnetic interference, special measures should be taken with respect to grounds, power, and layout. Pure symmetry is required for redundant block symmetry. Such precautions are already the expectations of automotive customers of semiconductor devices in automotive practice, for functional reasons. The guidance discusses the dependent-failure advantages of such symmetry and advises that diverse layouts do not always improve common-cause performance on semiconductors, but diverse techniques improve common-cause performance. This guidance is accepted but not anticipated by automotive semiconductor customers. Such layout diversity is used in non-semiconductor circuits for improved common-cause performance. This guidance helps harmonize automotive understanding, and semiconductor suppliers reference it.

ISO 26262-11 suggests approaches to verify the architectural metrics of semiconductor integrated circuits. In automotive practice, such verification can be time-consuming and resource-intensive, depending on the complexity of the circuit and the depth of the verification. One approach suggested is expert judgment based on the extent and validity of the rationale for the data. There is an example where 100% coverage is claimed for voltage regulator diagnostic coverage of over- and undervoltage faults of sufficient duration to cause failures. Some experts in automotive are hesitant to agree with 100% diagnostic coverage, while others accept the rationale. An argument is also expected to provide evidence and a rationale that the underlying fault that triggered the undervoltage did not also trigger a fault elsewhere on the semiconductor circuit. Convincing dependent-failure arguments encourage this guidance to be accepted. Automotive practice expects such rationales, and the metrics are then accepted.

Lists and descriptions are provided in ISO 26262-11 of safety mechanisms for analog integrated circuits. Such safety mechanisms are accepted in automotive practice, and the descriptions are generally agreed on. Effectiveness needs to be shown in order to determine the diagnostic coverage that is acceptable in automotive practice. Often this requires justification similar to that shown in ISO 26262-10. Customers may audit diagnostic coverage rationales, and this documentation should be available.

The ISO 26262-11 guidance for avoiding systematic errors in analog circuits consists of a list of measures that are applied during different phases of the development lifecycle, including entries for requirements management using tools, a heavy emphasis on simulation tools during design, and tool-based design verification. Prototypes are recommended to supplement design reviews. Requirements traceability to verification, testing, and management of special characteristics is also listed. These methods are expected to be broadly accepted in automotive practice because they mirror similar methods already in use at VMs, tier 1 suppliers, as well as semiconductor suppliers.

Safety documentation guidance for analog and mixed-signal semiconductor components is included in ISO 26262-11. It is advised that most analog integrated safety-related semiconductor components are developed in a distributed development. It is important that the customer develop the specification and that it is understood correctly by the supplier. The remainder of the documentation is per the hardware development lifecycle and DIA, which is as expected in automotive practice; much occurred before ISO 26262 was published.

PLDs

Guidance concerning functional safety of PLDs is provided in ISO 26262-11 in terms of tailoring the safety lifecycle requirements of ISO 26262 with respect to a PLD developed as an SEooC. The requirements are very much like the requirements for other semiconductor devices and are expected to be generally agreed on in automotive practice. Safety managers are required for the PLD manufacturer as well as the item developer. Assumptions must be considered by the item developer: if no safety mechanisms are provided internally by the PLD manufacturer, then only failure modes and rates are provided for the item developer. Examples of PLD failure modes are provided, and these are referenced in the automotive industry. Dependent-failure analysis as well as the guidance on failure-rate calculation are consistent with the expectations of automotive practice. There is also guidance for PLD users to avoid systematic faults that are referenced by auditors when auditing safety-related applications using a PLD. The documentation for a PLD application primarily concerns process, analyses including DFA, safety mechanisms, and assumptions. The DIA defines the work products, and they are jointly agreed to.

Multi-Core Devices

ISO 26262-11 provides guidance on the application of ISO 26262 to multi-core devices. Many hints are provided concerning freedom from interference by way of examples and notes. When these are implemented in the multi-core device, it is expected that they will add credibility to the freedom from interference claimed for the device. For example, it is suggested that virtualization technologies employed when using a hypervisor may guarantee freedom from interference caused by software elements. Measures to avoid timing errors are suggested, which is expected practice.

Sensors and Transducers

Sensors and transducers are discussed in ISO 26262-11. The distinction is made between a transducer that converts energy from one form to another, like a micro electromechanical system (MEMS), and a sensor that includes the transducer and provides a useful input for an electrical system. This is understood in the automotive industry, and although not everyone is rigorous in the use of the terms, all agree that this distinction is correct; this ISO 26262-11 guidance will help make the distinction more widely employed in automotive practice. For example, when a safety engineer from a tier 1 supplier meets with a safety engineer from a sensor supplier, the dialogue is expected to have improved precision as a result of the global discussions among the diverse safety experts who drafted, commented, and agreed on the guidance in ISO 26262-11 concerning sensors and transducers. Such discussions were not observed by this author during the six years preceding the first edition of ISO 26262, perhaps because there was not the assertive participation of tier 2 suppliers in the drafting. It can be expected that the automotive industry will continue in this direction while adopting the second edition of ISO 26262.

Failure modes of transducers and sensors are also discussed in ISO 26262-11 in some detail. The examples provided, though not exhaustive, should be checked in audits of sensors to ensure that they have been considered. Likewise, an example concerning cameras provides guidance for safety reviews not only in functional safety, but also in SOTIF. The increased emphasis on sensor failure modes and limitations will increase the emphasis on this guidance in ISO 26262-11. The guidance includes information concerning production processes and failure modes of MEMS. The focus is primarily on failures induced – for example, via mechanical stress – by processes of both the MEMS supplier and the MEMS customer. The causes and failure modes, while not exhaustive, are a more comprehensive collection than previously standardized for both MEMS customer and suppliers in the automotive industry. Automotive practice can now address these causes, which are referenced regularly.

MEMS

Failure-rate determination guidance is provided for semiconductor MEMS in ISO 26262-11. It is a comparative approach using an established baseline to establish an estimate for a new technology when no field data exists. The reasoning is expected to be accepted in automotive practice. The discussion of DFA provides a list of DFIs and examples that are used and checked in automotive practice. Many additional examples, such as packaging, can be added to the examples, depending on the application. The list of classes will trigger further discussion as dependent failure is scrutinized.

The guidance in ISO 26262-11 also includes advice for including MEMS transducers in the quantitative analyses of ISO 26262-5. The level of granularity is listed as a consideration. Automotive practice is to rely on the MEMS supplier for guidance while reviewing the DFA and safety mechanisms for supporting evidence. A non-exhaustive list of safety mechanisms provided is referenced in automotive practice. Coverage is determined by referenced techniques, and use of these techniques may evolve in actual automotive practice, just as automotive practice may evolve from using these techniques. Dedicated methods are also referenced and document existing practices. ISO 26262-11 provides additional elaboration on applying methods to avoid systematic errors concerning sensors and transducers; these measures agree with automotive practice. Recommended documentation for sensors and transducers supports automotive expectations and agrees with digital and analog documentation. The examples can be referenced to support compliance with ISO 26262.

Other Examples

ISO 26262-11 provides an extensive example of an evaluation of a DMA safety mechanism. The example breaks down the failure modes based on a granularity consistent with the guidance, and it will probably be accepted and referenced as long as there are not unresolved dependent-failure concerns. The evaluation also provides examples of quantification of the coverage of failure modes and other estimates. The methods employed as well as the actual estimates are accepted automotive practice. The techniques employed are widely used, so general acceptance is expected.

Digital and analog examples of DFA are provided in ISO 26262, and there is enough detailed information for these examples to be useful in automotive practice. In the digital analysis, the functions and blocks are explained at a fairly high level of abstraction to establish relationships for use in DFA. Then the safety requirements are derived, analyzed with an FTA, and examined with respect to DFIs. The example is consistent with the guidance in ISO 26262 and is expected to be referenced in automotive practice when similar analyses are reviewed. Similar enhancements are included. Likewise, the analog example is described at a functional block level so the safety requirement can be analyzed by an FTA. The shared regulator and identified coupling factors are examined, and mitigations are identified based on these coupling mechanisms. Such an approach is expected in the automotive industry, along with additional coupling factors and a justification for completeness. The example can be used for reference, and general acceptance is expected.

ISO 26262-11 includes examples of quantitative analysis for components. The digital example is somewhat restricted; nevertheless, it demonstrates the intent of the guidance provided. The notes are referenced in automotive practice for recommendations for performance of the SPFM. The format for handling permanent and transient faults is broadly adopted. For the analog component's example, the requirements are first determined to enable the analysis, after which violations and safe failures can be determined. The SPFM is performed, and safety mechanisms are determined. The notes indicate the necessary steps and order of execution; this sequence is accepted as automotive practice, but there may be some modifications, and safety mechanisms are assumed. The examples of quantitative analysis for PLD components first document the assumptions of use of the PLD, the derived requirements, and failure modes. Safety mechanisms inside and outside the PLD are addressed, as well as the failure distribution. The approach is expected to be an acceptable automotive practice when combined with a DFA – the DFA is not included, but automotive practice requires it.

ISO 26262-12 – Adaptation for Motorcycles

Background

During the preparation of the first edition of ISO 26262, there was participation by companies involved in the motorcycle industry. However, application to motorcycles was out of scope of the first edition. The motorcycle industry then developed a PAS to provide guidance to support the use of the first edition of ISO 26262 for implementing functional safety in motorcycle applications. In the second edition, the scope was broadened to include motorcycles. ISO 26262-12 provides guidance and requirements for motorcycles by including the needed tailoring of the other parts of the second edition of ISO 26262, based on the PAS. The increased maturity gained from using the PAS supports broad acceptance in the motorcycle industry.

The requirements for establishing a safety culture are consistent with requirements contained in ISO 26262-2, including communication with cybersecurity experts, and support the cybersecurity communication requirement added to ISO 26262 second edition. This is expected to receive broad industry support. The guidance for independence required for confirmation measures, while similar to the independence requirements in ISO 26262-2, was revised for motorcycles. These revisions were initiated by a consensus of the functional safety experts for the motorcycle industry, and approved for standardization, so broad acceptance is expected. The explanation for I3 independence makes it clear that a single enterprise can internally fulfill I3 requirements; this provides additional clarity for the ISO 26262 requirements, and consistent practice is expected.

Hazard and Risk Analysis

Perhaps the most significant information provided in ISO 26262-11 is the discussion of hazard and risk analysis. While the classification of *severity* and *exposure* does not differ from ISO 26262-3, the additional clarification of how to determine severity and exposure adds clarity that is broadly used as a rationale in actual practice: for example, when a feature fails to function. It is simple and concise, and tailoring for controllability is similar. There is guidance concerning the actions of the rider as well as other participants. Without this guidance for consistency of assumptions, there could be broad variations in the determination of controllability.

The greatest impact on tailoring is in the determination of the MSIL; this determines the index to be used to determine which requirements of ISO 26262 will be considered relevant for avoidance of a particular hazard. The mapping of MSIL by reducing by one ASIL is expected to be accepted broadly in practice. Safety goals then inherit this ASIL, which reduces the level of rigor to obtain an acceptable level of risk for motorcycle applications. ISO 26262-12 can be referenced for a rationale, and auditors accept this rationale.

Verification and Validation

The tailoring of verification and validation requirements for motorcycles in ISO 26262-12 provides clarification that may be particularly useful for applying methods and measures consistent with ISO 26262 to motorcycles. Many of the useful hints are in notes, which are not normative but are expected to be widely accepted in practice. They can be used as rationale, which is accepted by auditors.

The motorcycle workflow provided in ISO 26262-12 is consistent with the workflows of the other parts of ISO 26262. This workflow is widely used for planning and assessing motorcycle functional safety processes. An explanation of HARA for motorcycles is provided by ISO 26262-12 and is consistent with automotive practice to comply with ISO 26262. Significant clarification for motorcycle applications is gained from the explanations and tables for severity, exposure, and controllability that are specifically for motorcycles. The severity examples describe falls and collisions of motorcycles that can be used to calibrate the determination of severity in a hazard and risk assessment in motorcycle applications. Likewise, the probability of exposure examples for both frequency and exposure-determination methods are specifically for motorcycles. Examples of controllability of potential hazards are specifically for situations that are encountered by motorcycles. Determination is by riders considered to be experts by the enterprise. Guidance is provided for techniques to evaluate controllability, and it is expected that these examples, though informative, will be treated as a benchmark for credibility. This has been common automotive experience, and similar motorcycle adoption is expected.

References

1 Leveson, N.C. (2012). *Engineering a Safer World*. Boston: MIT Press. Print.

2 International Electrotechnical Commission (2010). Functional safety of electrical/electronic/programmable electronic safety-related systems. IEC 61508.

3 International Standards Organization (2011). Road vehicle—functional safety. ISO 26262.

4 International Standards Organization (2018). Road vehicles – safety of the intended functionality. ISO PAS 21448.

5 Miller, J. (2016). Chapter 5, The business of safety. In: *Handbook of System Safety and Security*, 1e (ed. E. Griffor), 83–96. Elsevier.

6 SAE International 2015. Considerations for ISO 26262 ASIL Hazard Classification. SAE J2980.

7 Machiavelli, N. (1961). *The Prince* (trans. G. Bull). London: Penguin. Print.

8 Miller, G.A. (1956). The magical number seven, plus or minus two: some limits on our capacity for processing information. *Psychological Review* 63: 81–97.

9 RESPONSE 3 (2009). Code of practice for the design and evaluation of ADAS. Integrated Project PReVENT. https://www.acea.be/uploads/publications/20090831_Code_of_Practice_ADAS.pdf.

10 United States Department of Defense (1949). Procedures for performing a failure mode effect and critical analysis. MIL-P-1629.

11 National Aeronautics and Space Administration (NASA) (1966). Procedure for failure mode, effects, and criticality analysis. RA-006—13-1A.

12 AIAG (1993). *Potential Failure Mode and Effect Analysis*. Automotive Industry Action Group.

13 AIAG (2008). *Potential Failure Mode and Effect Analysis (FMEA)*, 4e. Automotive Industry Action Group.

14 SAE International (1994). *Potential Failure Mode and Effects Analysis in Design (Design FMEA), Potential Failure Mode and Effects Analysis in Manufacturing and Assembly Processes (Process FMEA), and Potential Failure Mode and Effects Analysis for Machinery (Machinery FMEA)*. SAE J1739.

15 SAE International (2009). *Potential Failure Mode and Effects Analysis in Design (Design FMEA), Potential Failure Mode and Effects Analysis in Manufacturing and Assembly Processes (Process FMEA)*. SAE J1739.

16 Automotive SIG (2010). Automotive SPICE Process Reference Model.

Automotive System Safety: Critical Considerations for Engineering and Effective Management,
First Edition. Joseph D. Miller.

Index

Automotive System Safety: Critical Considerations for Engineering and Effective Management, First Edition. Joseph D. Miller.

Printed and bound by CPI Group (UK) Ltd, Croydon, CR0 4YY
16/04/2025
14658391-0003